The Biophysics Basis for Acupuncture and Health

by Shui Yin Lo, Ph.D.

Dragon Eye Press
Pasadena, California

DRAGON EYE PRESS
Pasadena, California

Copyright © 2004 by Shui Yin Lo

All rights reserved under International and Pan-American Copyright Conventions. Published in the United States by Dragon Eye Press.

Library of Congress Cataloging-in-Publication data

Lo, Shui Yin.
The Biophysics Basis for Acupuncture and Health/ Shui Yin Lo. – 1st ed.

ISBN 0-9748261-0-3
1. Health – Acupuncture   2. Science – Quantum Physics   3. Biophysics   4. Medicine –Acupuncture   5. Physics   6. Asian Medicine – Acupuncture   7. Title

Printed in the United States of America

CONTENTS

Acknowledgments   xi

Foreward   xiii

Introduction   xv

About the Author   xvii

Preface   xix

Chapter 1. Meridians – A Network of Stable Water Clusters

1.1. A universe within a universe.   1
   The outer multi-layered universe can be explained with quantum theory. The inner universe is waiting to be explained by the same theory. Meridians and quantum: a dual analysis with taichi diagrams. The unification of quantum theory and meridian theory; theory of electrons, ions and photons; quantum theory and its universality of application, from universe, to human being, to quark.

1.2. The molecular structure of meridians.   11
   Vibration of water clusters in meridians. Water clusters: meridians are hypothesized to be made up of polarized molecules; signals are transmitted mechanically, with photons, electrically and/or with ions. Evidence favoring water clusters as constituents of meridians.

1.3. Acupoints and acupuncture.   18
   Acupoints as local organizing centers; acupuncture – inducing natural oscillations by whatever means. Stimuli at acupoints: mechanical, electrical, photonic and with material injection.

1.4. Quantum theory, predictability and uncertainty in life.   26
   Role of quantum theory in understanding life and acupuncture. A human being is described by wave functions. His activities are described by probability amplitudes: probabilistic, initial and final states. Quantum numbers – discrete states. Crowd effect (symmetric wave function); exclusivity (anti-symmetric wave function). Feynman space-time approach as necessary to holistic description of human beings and oneness with the universe.

Chapter 2. Quantum of Life – Life as Vibration   33

2.1. Three levels of existence
Who am I? When organs vibrate, they are alive. When meridians vibrate, a person is alive. Three levels of meridians are discussed in a simplified model. The lowest level is their physical constitution, as in any inanimate object. The second level is the ground state of meridians in a living human. The third level is the excited level of meridians in a living human. The qigong state and higher levels of oscillations. Life is described by quanta of energy.

2.2. From qi to qions – oscillations as quanta of life   37
Sound as a particular kind of oscillation – piano, ear, brain, memory, musical notes.

2.3. Life - Creation of qions. Death – destruction of qions   41
Creation and annihilation are natural to living phenomena. Action principle as a coordination of all organs and systems.

2.4. Meridians – the primitive site of life   44
Various effects of ST 36 meridian; meridian, most primitive, and primary control system; origin of life; water clusters; relations with other systems in the human; auricular acupuncture; plant meridians; interactions among meridians.

2.5. Clicking the mouse   50
Computer as analogous to how meridians are related to the brain. Decision-making, balance. Three levels of existence. The role of the boson field; qion exchanges. Mother-son relationship among meridians.

Chapter 3. Acupuncture – The Meridian System as the Most Fundamental

3.1. Meridians and survival   61
Acupuncture can affect the immune system, reproductive system, nervous system, gastrointestinal system, urinary system, stomach, spleen and liver.

3.2. Getting Well – what acupuncture cures and how it cures   67
Cardiac disorders, arthritis, asthma, cancer, diabetes and pain.

3.3. Symmetry in human beings   76

3.4. How do you know you are sick? – diagnosis   78
Methods of diagnosis; pulse method; electrical signals; infrared: direct and indirect.

3.5. Getting rid of pain exponentially – evidence from infrared imaging   80
An exponential law. Acupuncture initiates a healing mechanism that oscillates cells in a random statistical fashion.

3.6. What acupuncture cannot do Direct and indirect approaches: integrating East and West.	81

Chapter 4. From Vibration to Vibrant Health

4.1. To become healthier Obesity; aging.	83
4.3. Resonance as a mechanism for acupuncture If blockage occurs and a meridian does not vibrate at normal amplitude and frequency, we feel pain. Auricular acupuncture. Meridians as a network. Helping others by emitting qi; its evidence and its quantum fields.	87
4.4. Vibrant health What can a person with vibrant health do?	90
a. Radiation from qigong masters.	
b. Orderly alignment of water clusters in qigong masters.	
c. Diagnoses with external qi. Interacting with other persons.	
d. Polarization of infrared radiation.	

Chapter 5. Standing on Solid Ground – Philosophy and Mathematical Formulations

5.1. Taichi diagrams – a dual analysis of opposites. Chinese view of life: change and transformation as the only constants. Consistent with quantum theory. Taichi illustrates the idea of the unity of opposites. Mathematical formulations can express this as quantum theory.	95
5.2. Quantum fields for bosons.	98
5.3. Electrons, fermionic qions and symmetry.	100
5.4. Dramatic approximation for a human being.	102
5.5. Survival, family, and gauge principle.	108
5.6. The action principle, interaction, and Feynman diagrams.	111
5.7. Dipole radiation; exponential law in pain relief.	115
Water Circles – a poem	120
References	121
Pictures	143
Tables – Clinical Research on the Effects of Acupuncture, Acupressure and Moxibustion	161

1.2.1. Meridians as optical fiber - optical properties of meridians	163
1.2.2. Meridians as conductors - evidence for low impedance along the meridians	165
1.2.3. Different effects from different frequencies of electro-acupuncture	167
1.2.3a. Summary of differences between 2 Hz and 100 Hz in electro-acupuncture	171
1.2.4. Meridians as plastic tubes with water - propagation of sensation along meridians (PSM)	172
1.2.5 Constituents inside meridians are charged - evidence for meridians: ionic and other	174
1.3.1. Properties of acupoints – acupoints as water well, or as local organizing centers	177
1.3.2. Effect of laser acupuncture; photonic stimulation of meridians	181
1.3.3. Effect of transcutaneous electric nerve stimulation at acupoints	183
1.3.4. Evidence of acupressure – mechanical stimulation of acupoints	187
1.3.5. Evidence of injection at acupoints – stimulation of acupoints with foreign substance	189
1.3.6. Evidence of moxibustion – meridians as infra-red transparent optical fiber	190
2.3.1. Meridians in plants – meridians as most primitive and fundamental system	195
2.3.2. Various effects of stimulating ST 36 zusanli	197
3.1.1. Effect of acupuncture on immune system	201
3.1.2. Effect of acupuncture on nervous system	202
3.1.3. Effect of acupuncture on reproductive system	205
3.1.4. Effect of acupuncture on gastrointestinal system	210
3.1.5. Effect of acupuncture on urinary tract	216
3.1.6. Effect of acupuncture on the liver	219
3.1.7. Effect of acupuncture on spleen	220
3.1.8. Effect of acupuncture on stomach	222
3.1.9. Effect of acupuncture on lung	224
3.2.1. Effect of acupuncture on cardiac disorders	225
3.2.2. Effect of acupuncture on arthritis	231
3.2.3. Effect of acupuncture on allergies	236

3.2.4. Effect of acupuncture on cancer patients	240
3.2.5. Effect of acupuncture on diabetes mellitus	249
3.2.6. Effect of acupuncture on headache	252
3.2.7. Effect of acupuncture on neck pain	254
3.2.8. Effect of acupuncture on shoulder pain	255
3.2.9. Effect of acupuncture on low back pain	258
3.3.1. Symmetry effect of acupuncture	261
3.5.1. Exponential decay	262
4.1.1. Effect of acupuncture on obesity	263
4.1.2. Effect of acupuncture on aging	269
4.2.1. Effect of qigong on healthy persons	271
4.2.2. Effect of qigong therapy	274
4.3.1. Effect of auricular acupuncture	277
4.3.2. Evidence against nerves and for meridians as responsible for effects of acupuncture	281
4.4.1. Effect of external qi on animals	283
4.4.2. Physical properties of external qi by qigong masters	284
5.5.1. Conservation of number laws	285

Acknowledgments

I wish to thank all the acupuncturists that have worked with me. In particular, Kuo-ching Lo, the first one to work with me on infrared imaging in the field of acupuncture, and who explained to me the many intrinsic difficulties in using modern scientific terms to describe Oriental medicine. Roger Chen, Senlin Wang, Jing-ru Zhou, L.C. Li, L.C. Wen, and Bridget Cheng have all worked with me to obtain much valuable data. I'd also like to thank the many acupuncturists and scientists who attended my lectures and asked many probing and relevant questions that helped to shape the format and content of this book.

I also wish to thank Andrew Chiu, Cedric Ling, James Shui-Ip Lo and Ben Shui-chun Lo for their valuable comments.

I owe gratitude to my family: Angela, my wife, who was supportive even to the last day when she passed away; my daughter Fiona Ai-ming; and my son Haomin, for their support of my persistent effort for many years. My son, Alpha Wei-min Lo, is especially thanked for his relentlessly critical remarks on the validity of physics in biological systems and for his beautiful cover design.

Karen McChrystal, MA, my editor, has contributed enormously to the final appearance of the book. She worked extensively and diligently beyond the call of duty on all aspects of fine-tuning the present book to make it readable for a general audience.

Foreword

Historically, when paradigms are shaken and come into contact with other paradigms, new and powerful things emerge. This book is about a subject that is simultaneously at the intersection of several paradigms – that of medicine and science, the reductionist view versus a systems view, the Newtonian classical view versus a quantum view, Eastern medicine and Western medicine, the physical being and the non-physical.

When medical procedure is accompanied by an underlying knowledge of science, it can be a powerful marriage. Western medicine has a basis in Pasteur's germ theory, Harvey's blood circulatory theory and twentieth century molecular biology. This book attempts to provide a modern basis for Oriental medicine by providing an underlying biological and physical framework for meridians and qi, and thus a Western scientific basis for acupuncture, moxibustion and qigong. Such a framework is of fundamental importance.

The reductionist view holds that we can understand anything in terms of its building blocks. This has been the approach of science for some time – one breaks everything into its parts and tries to understand the thing from its parts. In the twentieth century, though, a more whole systems view of science has begun to develop, including quantum field theory, nonlinear dynamics, systems theory, self-organizing systems, fractals and network theory. These all give importance to the relationships among parts, i.e., how things cohere within a whole.

To illustrate the difference between these viewpoints, take the example of a workplace. Say one person gets overloaded with work and becomes sick. The reductionist viewpoint looks at the basic building block, the person, and says there is something wrong with that person. A systems scientist might say that everyone may be passing all his or her work to one person, instead of distributing it. To rectify the problem, the reductionist might hire a worker who can work faster. A systems scientist would redistribute the workflow.

Oriental medicine is a systems science in that it looks at the relationships among the different parts of the body. The body is evaluated as being in harmony or not. Meridians are the fundamental system of the body, governing the whole. The circuit of meridians connects the entirety of the body, so that a needle inserted at one point can affect a point distant from the needle. This book proposes that there are oscillations in the meridians of some polarized medium, which is most likely made up of stable water clusters with permanent electric dipole moments, that resonate with sympathetic oscillations in an organ or system to bring it into balance or harmony with the rest of the body. Health consists in keeping the body in tune with the natural oscillation frequencies of the body.

Meridians and acupoints form a communication system, and so can create harmony in the body. To use the workplace analogy again, if a group of people are not working well together, from a systems viewpoint it may be that they are not communicating with each other, rather than that there is something wrong with one subgroup of the people.

A systems viewpoint would look at how energy flows through the system. Qi flows through the system of meridians. A systems viewpoint significantly deepens the way we approach the body, and helps us understand medicine from a more comprehensive and integral point of view.

In the beginning of the twentieth century, quantum theory hit the world and shook it up. Before that, in the Newtonian worldview, things were seen as solid objects, and one could calculate, as with billiard balls on a table, how everything would evolve deterministically. Quantum mechanics says that the underlying reality is made up of fluctuations – things are not necessarily in one place, things are not still, and things can be created and annihilated out of an underlying vacuum. The behavior of objects is more probabilistic than deterministic. Quantum mechanics changes the way we look at the world philosophically.

The mainstream view of the body in biology has been pretty much that of the deterministic classical Newtonian model. This book attempts to set forth a framework for looking at the body from the viewpoint of quantum mechanics. The body is made of quantum vibrations. There are quantum oscillations of meridians and of qi all the time. This is the basis for vibrant health.

A lot of quantum affects are not readily apparent by the time we get to the large human scale, but if they were still to be there, it would explain the basis for vibrant health as well as some extraordinary behavior. It is postulated that one such affect would be that, with ions flowing through them, meridians would become superconducting at some state.

Just as understanding the world from a quantum viewpoint caused a revolution in physics, understanding the physical body from a quantum viewpoint can cause a revolution in how we see the body and health. The Oriental model does not contradict the Western model, wherein the body is analyzed in terms of cells, organs, nervous system, blood circulatory system, immune system, and so on. It proposes another system existing in the body, the meridian system, which connects with and interacts with all the other systems on a fundamental level. Oriental medicine has been speaking a somewhat different language than Western medical professionals have been used to, a language of qi, meridians and harmonizing the body. By proposing an underlying biophysical mechanism, and by looking at Oriental medicine from systems science, it is hoped this book can help in laying a path for the integration of Western and Oriental medicine.

Alpha Lo

Introduction

What are meridians? What is qi? How does acupuncture heal? Why does the insertion of a needle trigger a whole set of biochemical reactions in the body? What is it that keeps us healthy?

This book answers these questions by bringing together quantum theory and massive scientific clinical research evidence on the effects of acupuncture, from various countries, including the UK, Sweden, Germany, Japan, the USA and China. It looks at the body from various points of view.

From the clinical point of view, this book summarizes massive scientific evidence from experiments and research on the effects of acupuncture, acupressure, moxibustion and qigong. Acupuncture balances the body by stimulating the body's natural functioning and healing processes. This occurs at all levels, including biological tissues, t-cells, neuronal firing, organs, endorphins, blood circulation and body temperature. On the systems level, acupuncture affects the immune system, nervous system, gastrointestinal system, urinary system et. al. It also works on the level of organs, including the liver, spleen, stomach, lung and heart. Acupuncture is beneficial in healing a wide array of diseases, including heart disease, arthritis, allergies, cancer, diabetes, headache, neck pain, shoulder pain, low pack pain, obesity and aging.

From the viewpoint of molecular biology, it is proposed that meridians are electrical structures, comprised of chains of polarized water clusters, a kind of structured water different from normal water. When these clusters are aligned, qi flows smoothly, unimpeded, and we are healthy. When water clusters fall out of alignment, qi is stagnant and we feel pain or become ill.

From a physics viewpoint, the book explains how sound waves, electromagnetic radiation and ions flow along the meridians, transmitting information and energy. The ion flow around the whole meridian circuit corresponds to the 24-hour oscillation of qi.

The book discusses how stimulations at an acupoint affect the oscillations in an organ, which then affects the physical body at many levels, including the infrared radiation from different parts of the body. Infrared light propagates along meridians and outside of them. This is visible in infrared photography. Parts of the body surface that are hotter than normal indicate either where there is pain or illness, or where activity is occurring to heal the problem. The author includes infrared photos he took of patients receiving moxibustion therapy. They show conditions before and after treatment, providing a useful methodology for acupuncturists, qigong practitioners and other health practitioners. Other possible diagnostic tools are also discussed.

From an information processing viewpoint, meridians are seen as the underlying system, governing all the other systems of the body, including the immune system, the reproductive system, nervous system, gastrointestinal system and urinary system.

The book discusses how the meridian system transmits signals to different parts of the body. These signals could be sent in the form of electromagnetic signals, phonons or ionic flow. The acupoints are the local organizing centers of information flow, behaving like a microprocessor.

In the quantum physics view, the body may be seen as a probabilistic wave function. The body and meridians exist in discrete quantum states. There are three basic states of life – the dead state, when the structure is there but nothing is happening; the ground state, when things are just alive; the excited state, when things are alive with energy.

This quantum viewpoint looks at the oscillations of water molecules, organs, meridians and ions as quantum oscillations. These quantum oscillations are related to energy flow in the body, information flow, and correlate with different types of healing. The oscillation of the water clusters can resonate with the oscillations in other parts of the body, thus harmonizing them. According to the author, these quantum oscillations may give rise to coherent radiation. This radiation may also be polarized. And it is possible that the wave function becomes macroscopic, giving rise, perhaps, to superconductivity.

From the viewpoint of quantum field theory, qions, the quantum language for a particle of qi, are posited as the quantum field of qi. Qions can be annihilated or created, absorbed or released by water clusters. The interaction of meridians is described as the interactions among qions.

The effects of qi and qigong on health are described and validated by the research collected by the author. Qions are posited. These are created and annihilated, may be absorbed by other particles, and they travel over all space and time.

The effect of qigong is described in terms of the emission of qions that travel over all time and space. This may explain, among other things, how a qigong master is able to heal people by emitting external qi. There seems to be a relationship between external qi and coherent infrared light.

In combining these different perspectives, the author hopes to have presented a comprehensive view of the underlying mechanisms and dynamics that lead to ill health and, moreover, a view of the underlying basis for vibrant health.

Alpha Lo

About the Author

Dr. Shui Yin Lo is a Professor of Chinese Medicine, former Professor and international lecturer in the areas of Acupuncture, Theoretical Physics, Quantum Field Theory, Particle Physics, Mathematical Physics, General Relativity, Electromagnetic Theory, Classical Mechanics, Thermodynamics, Advanced Quantum Mechanics, Advanced Electricity and Magnetism, Statistical Mechanics, and Particle Physics. He has supervised four Ph.D. and eight Masters Degree students in high energy physics.

He is the author of a number of publications, including four books. He also has a number of patented inventions, including medical devices; infrared imaging equipment; a system for nuclear fusion generation; non-polluting, combustion-enhancing fuel additives; an emission control device; a home water distiller, and numerous other patented inventions in the field of quantum physics.

Dr. Lo received his Ph.D. in Theoretical Particle Physics from the University of Chicago, was visiting faculty at the California Institute of Technology for many years, and is now residing in Pasadena, California, USA.

Dr. Lo may be contacted by email at ideaclinic@yahoo.com.

Preface

This book addresses the biophysical basis for acupuncture. Orthodox science thus far has no acceptable theory of meridians and why acupuncture works. We propose a molecular structure for meridians and a quantum theory to explain the effect of acupuncture.

There are two pillars of the book. The first is the quantum physics developed in the twentieth century. The second is Oriental medicine developed from the *Canon of Medicine* of China, written two thousand years ago. Quantum physics is the pinnacle of modern science in its rigor and completeness.

One could easily argue that Oriental medicine has passed clinical tests of at least 100 million patients in its long history of treatment, following the same essential recipe prescribed in the *Canon of Medicine*. Otherwise it would not still be used, as my acupuncturist friends like to say. Modern Western drugs usually have to pass clinical tests involving hundreds or sometimes thousands of patients.

Statistical physics has two kinds of averages: ensemble average and time average. They should produce the same result and have the same rigor. The clinical results of Oriental medicine are statistically closer to the time average, and the clinical tests of modern Western drugs are statistically closer to the ensemble average.

In the past fifteen years there have been many studies all over the world using accepted Western scientific methods and apparatuses to test various aspects of meridians and acupuncture. These studies are generally written in highly technical papers that are usually understandable only to specialists. Ordinary laymen would not enjoy reading them, but nevertheless would like to know they are there to support the general arguments in this book. So we have compiled these clinical and experimental results in concise tables. Experts should read them, and ordinary people can skip them if they choose.

For ease of understanding, the text is written in question and answer format. Readers can choose to read anywhere in the book for any given topic. It is not necessary to read the entire book in sequential order.

Since the average person may have difficulty understanding mathematical equations and abstract physics theory, we've written the first four chapters without equations. We put all of the equations in the last chapter, which may be skipped by the average reader.

The book is written for patients, physicians, acupuncturists, scientists and ordinary persons. They all should read for the following reasons:

- Patients should read this book in order to learn about the scientific evidence for the effectiveness of acupuncture. The book describes the fundamental nature of the human body at the level of electrons and the movement of qi energy throughout the body, traveling through the meridians.

- Physicians should read this book in order to gain a scientific understanding of how acupuncture works and why. For example, it is explained how a single acupuncture needle can cure many diseases.

- Acupuncturists should read this book to help them explain to their patients in scientific terms how and why acupuncture works.

- Scientists should read this book to understand the human body in simple terms, in a more unified framework than Western medicine so far provides. The book shows how meridians are the fundamental system that governs the rest of the human body. This view is entirely holistic.

- Laypersons should read this book to understand more about the human body and how it is all interconnected. This book gives scientific evidence for the genuinely holistic nature of the body – the body works as a unified whole and the meridian system is the fundamental, primitive and unifying basis governing the entire person. It even underlies states of consciousness, from sleeping, to waking, to meditative, to active, to spiritual.

Body, mind and spirit are really one entity in different quantum states. Quantum theory explains the transitions between these quantum states.

Chapter 1

Meridians – A Network of Stable Water Clusters

A universe within a universe

The Chinese theory of medicine, as derived from the ancient Chinese medical classic, the *Canon of Medicine* (*Nei Jing*), regards the human being as a small universe, made up of atoms, molecules and multi-levels of matter. In the modern view, quantum theory, which governs the behavior of atoms and molecules, can be applied both to the multi-leveled universe and to a multi-leveled human being.

Q1.1.1. What does it mean that the universe is made up of multi-levels of matter?

The multi-leveled nature of the universe can be discussed in three different variables: space, time and energy.

The Space Variable

In space, the largest object is the universe itself, which is about one million trillion (10^{18}) meters. Inside the universe there are many galaxies, which are about ten thousand times smaller than the universe (10^{14} meters). Our earth is a small planet in our solar system, which is about ten million times smaller than a galaxy (10^7 meters). A human being is, on average, about 1 to 2 meters tall. Humans are made up of molecules, which are about one billion times smaller (10^{-9} meters). Molecules are made up of atoms, which have in their center atomic nuclei one million times smaller (10^{-15} meters).

Physics can explain the major features of all of these, from the very large to the very small. Quantum theory can explain the big bang, from which the present universe comes, the evolution of the galaxy, and all the way down to nuclear reactions. It is quite reasonable to expect that quantum theory will be able to explain the behavior of human beings. (See Fig. 1.1.1.)

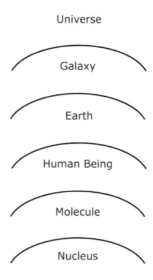

Fig. 1.1.1. Quantum physics can explain the very small, the very large, and everything in between.

The Time Variable

In the time variable, shown in Fig 1.1.2, the uranium nucleus has one of the longest lifetimes, which is about four billion years. On Earth, an ice age lasts about one hundred thousand years. The lifetime of a neutron is about ten minutes. The particles called mesons, which bind the atomic nuclei together, have a lifetime of about 10^{-23} seconds. In the time variable, the universe is composed of multi-leveled matter, each level having a different order of magnitude of lifetime. Quantum theory can explain the fundamental mechanism that produces the wide variation in lifetime of uranium, neutrons and mesons. Human beings have a lifetime of about a hundred years, in the middle of the scale of these lifetime variations. It is quite reasonable to expect that quantum theory can explain the fundamental mechanism of life in human beings.

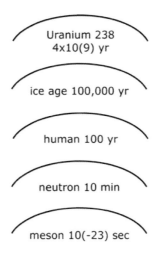

Fig. 1.1.2. Physics can explain the behavior of objects with a very short life, like mesons, a very long life, like uranium, and all those in between.

The Energy Scale

On the energy scale, the nuclear fusion that happens in the sun gives out enormous energy (3.6×10^{28} joules/sec). The nuclear fissions that occur inside the nuclear power reactor are much smaller, but can still give up to one billion joules per second. The thermal energy needed to raise one gram of water by one degree C is about 4.2 joules. The smallest lump sum of energy in an ultraviolet light is about one electron volt, or 10^{-19} joules. So the universe can be divided into levels of objects, each level having different energy scales. These range from very small (10^{-19} joules) to very large (10^{28} joules), as shown in Fig. 1.1.3. The energy output of a human being is about 10 joules per second, which is in the middle of the scale. Quantum theory can explain objects ranging from large to small in the energy scale. It is reasonable to expect that it can explain the energy scale of a human being.

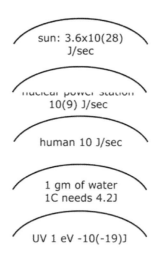

Fig. 1.1.3. Physics can explain the behavior of matter that is very energetic, like the sun, the very small, like UV photons, and everything in between.

Q1.1.2. Are human beings also made up of many levels, like the universe?

The universe has many levels. For every level there is a smaller level inside, like a set of Chinese boxes. The human body is also composed of many levels. There are organs, like the lung and the heart. Organs are made up of cells. Cells consist of molecules. Molecules are made up of electrons, nuclei and photons. (Photons are the smallest particles of electromagnetic waves.) (Fig. 1.1.4.)

Q1.1.3. Can you explain quantum theory in a few words and why we need it?

Quantum theory describes reality, at the microscopic level, as having coexisting particle and wave properties. The interaction of particles and waves is a way to describe the processes fundamental to life – creation, annihilation and the essence of life itself. So in applying quantum theory to a medical perspective, we do not look at separate entities, such as disease entities, but rather at the interactions of the parts within the whole. In essence, quantum theory lets us analyze a person's health in terms of his adaptability in relation to himself and to his environment.

Quantum theory has laws that govern the behavior of matter. These laws can predict future events. Future events, however, cannot be predicted with 100 percent accuracy. They occur with higher or lower probabilities. This is true for human events as well. The future of a human being is uncertain. Events in human life occur with differing probabilities. So quantum theory should be able to describe human events as well.

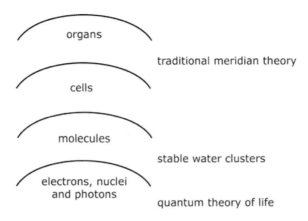

Fig. 1.1.4. Different layers of the human body and their relation to meridian theory.

In life sciences the central question is "What is life?" Life is created at birth. The process of creation is formulated precisely only in quantum theory. It can explain the creation of life, as much as it can explain the creation of light when we turn on the light switch in a room.

Q1.1.4. What is it that makes us alive, and different from the dead objects around us?

We have meridians, unlike dead objects. There are oscillations inside humans that are characteristic of life. A brain is alive because it has electrical oscillations in the brain waves. A heart is alive when the heart's mechanical system oscillates. A human being is alive if his meridians oscillate.

Q1.1.5. What are meridians?

It is common knowledge that the human body contains a nervous system and a blood circulatory system. According to ancient Chinese medicine, there is also a meridian system. In fact, according to the Chinese, the meridian system is more basic than the nervous system and blood circulatory system, especially in relation to a person's overall health.

Q1.1.6. Why do we not hear about meridians from our medical experts?

Meridians are not taught in orthodox medical texts. Perhaps the main reason is that in anatomy we find nerves and blood vessels, but we do not find any sign of a meridian. Our explanation is that meridians are made of matter, similar to what the surrounding tissues are made of, with the exception that meridians are ordered with a system of permanent electric dipoles.

Q1.1.7. What is qi?

Qi occupies an important place in Oriental medicine, but so far it lacks a modern scientific explanation. Our explanation is that qi is composed of oscillations of matter in the meridians. This is analogous to when we hit a steel pipe, atoms in the pipe oscillate back and forth, and a sound wave is generated. The oscillation of atoms results in a sound wave. Generating sound waves in steel doesn't add any weight or matter. Sound waves have some energy, but not that much. We use them mainly to carry information. We think that oscillations of matter along the meridians are similar to sound waves in steel pipe. They carry some energy, but not that much. We use them mainly to carry information.

Q1.1.8. How did you come to study meridians?

I am a theoretical physicist. I came to the field of meridians and qi fifteen years ago, when I went to learn qigong from a qigong master. After a few lessons, she came to know me and said that I should use my knowledge to understand qigong in a scientific way.

As I started searching through the literature on qi, meridians, acupuncture and qigong, the most startling feature I found was that all the authority behind the procedures and methods of acupuncture and meridians was found in an ancient book compiled in China, more than two thousand years ago: the *Canon of Medicine*, attributed to the Yellow Emperor. This was quite contrary to my upbringing in science. In modern science, authority goes back to experimental findings and not to an individual or to a book.

We may even say that the *Canon of Medicine* is probably the oldest book still being quoted, read and used in teaching any scientific discipline in the East.

Q1.1.9. Can you think of another example similar to the Canon of Medicine in the field of science?

The closest rival in antiquity to claim this degree of authority in science that I can think of is Euclid's Geometry. In plane geometry all the theorems discussed in Euclid's work are still valid. Even more important, the trend of thought laid out in this work determined the future course of development of mathematics in general. The idea that one has to prove any theorem in order for it to be valid and accepted comes from this work. A mathematician has to prove any result that he claims as new, or his fellow mathematicians will treat it as not being solid. All proofs eventually come from a few axioms. This axiomatic approach is the orthodox way of proving any mathematical theorem. Euclid's Geometry provides a standard for any branch of mathematics to follow.

Similarly, in the future, when we succeed in putting Oriental medicine on a solid scientific foundation, the philosophy of the Canon of Medicine would still be our guiding light on the subject of health.

Q1.1.10. What is the most sophisticated scientific theory of matter developed in the last hundred years? How does it relate to meridians?

After the first developmental stages of science, occurring during the glorious period of ancient Greek civilization, came the second pinnacle of scientific achievement – Newton's formulation of classical mechanics and gravitational force. In the twentieth century, I think, the highest intellectual theoretical development of the whole human race is the invention of quantum theory.

The purpose of this book is to set this old discipline of acupuncture and meridian theory in a modern scientific framework. The data we use will come mostly from scientific work and clinical trials done in the past fifteen years. The theory we rely on is currently the most sophisticated theoretical framework, the quantum theory (Fig. 1.1.5). Acupuncture and meridians are supported and verified through at least two thousand years and by millions of patients in clinical practice. The quantum theory is supported thoroughly by the most accurate, stringent, sophisticated experimental methods, apparatuses and results.

Readers should not be worried that they do not know what quantum theory is. We basically rely on quantum theory to provide the line of thought and the basic concepts to explain meridians and acupuncture. We shall explain these concepts as we go along.

Q1.1.11. How do you combine the most sophisticated quantum theory with the most ancient meridian theory?

In the same way that modern mathematicians accept the spirit of Euclidean geometry, we shall in this book accept the spirit of the ancient meridian theory as being essentially correct in describing the function of the human body and curing its diseases. Whenever we do not know how to proceed next, we shall appeal to this meridian theory for guidance and inspiration. Our task is to use concepts developed in quantum theory together with the latest findings in physics, chemistry, biology and clinical tests to explain the ancient meridian theory in modern scientific terms. Reinterpreting ancient theory within current scientific terminology will enable us to build a solid experimental foundation for a new meridian theory. The field of application of traditional Chinese medicine will then expand enormously and eventually be acceptable to everyone.

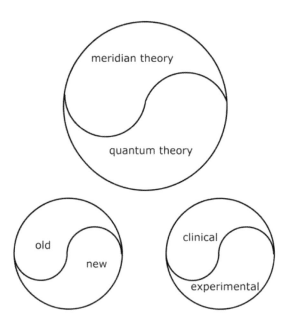

Fig. 1.1.5. Unification of meridian theory and quantum th

Q1.1.12. How do you explain traditional meridian theory using modern scientific understanding?

Traditional meridian theory generally concerns itself with the internal organs, Zhong-fu, which are the largest units just smaller than the human body itself. Anything smaller than internal organs are referred to in more abstract fashion, as qi, blood, wind, cold, etc. In ancient times nothing was known about cells and molecules. So it is natural that traditional meridian theory does not deal with them. Only in recent research, especially research with animals like rats and rabbits, is acupuncture found to be able to influence cells and molecules, which we shall discuss later. Clearly, a theory that deals with the life of human beings, their health and diseases, must explain all the different levels of the human body. Our hypothesis in using quantum theory to explain meridian theory is basically a theory of molecules, electrons, nuclei and photons. The new dynamic element we introduce here is the concept of qi, which will be explained as collective excitations, to use the language of quantum field theory.

From the viewpoint of orthodox Western medicine, cells are well understood. The primary focus of attention is on molecules. The latest and the most active research frontier is about the molecular composition of DNA. The complete mapping of DNA in humans is being done in the Human Genome Project. It costs billions of dollars, involves teams of excellent scientists, and is expected to yield incredible benefits in biological science.

DNA is the smallest molecular unit that is responsible for storing genetic information and passing it on from one generation to another. Presumably once the mapping of DNA is completed, there will be other important molecular questions to be answered, e.g., how do proteins work? To understand this, many variables must be accounted for. Any molecule generally has physical and chemical properties. The flow of blood and electrical nerve pulses are basically physical properties. The biochemical reactions of molecules, such as the burning of sugar and protein, are in the domain of chemical properties.

No one really knows how many different kinds of molecules comprise a human body. Even less is known about how many different kinds of biochemical reactions occur on the molecular levels. It's easy to measure all the variables and to see the results of the oxidation of carbon hydrate in a test tube. If we put carbon hydrate and oxygen in the test tube and heat it up, the resultant product is carbon dioxide and water. But inside the human body the burning of carbon hydrates proceeds at body temperature, through many steps, with the help of various kinds of enzymes. And it is very difficult to measure how it proceeds at each step. How can we sort out all these complex processes?

There is a way, at least in principle. If we study objects smaller than molecules and atoms, there is great simplification in concept. Millions of molecules and their trillion kinds of molecular interactions can be viewed simply as the movements of electrons, nuclei and photons. In quantum theory these are well studied and understood.

Q1.1.13. Why did you choose to study acupuncture in your application of a modern theory of meridians?

When I first looked at the effect of acupuncture with the eyes of a scientist, I was amazed by the simplicity of it. A single needle inserted in the body can cure hundreds of illnesses without medicine. No chemicals are involved. There are no chemical reactions introduced from outside materials. Yet the needle initiates all kinds of biochemical reactions inside the body.

How does it work? It must work at a more fundamental level than biochemistry. It must work on the level of electrons and nuclei, which exist inside the molecules and are bound together by an electric field (photons). So to simplify greatly, Western medicine at the moment concentrates on events occurring at the molecular level, whereas acupuncture works through electrical forces coming from the movement of electrons and nuclei. The commonality between Western and Oriental medicine may be found in the relationships between molecules and electrons.

In summary, the theory behind the practice of acupuncture is quantum theory. The needle exerts its effect on the human body by affecting the positive and negative charges. The nuclei bear the positive charge, which is yang. The electrons bear the negative charge, which is yin. The fundamental concept of yin and yang in Oriental medicine is to be explained in terms of negative and positive charges inside molecules of the human body.

Our explanation of traditional Chinese medicine uses yin and yang, whereby two polarities interact, with each side containing a small portion of the other. We integrate this explanation with the quantum theory of electromagnetism involving electrons, nuclei and their components.

Q1.1.14. What is the relation of this quantum meridian theory with the ancient meridian theory?

Let us come back to our analogy of Euclidean geometry. In the twentieth century, David Hilbert, one of the most important mathematicians, reformulated Euclidean geometry without the use of geometrical shapes at all. He used only modern notions of relations in mathematics, but these reproduced all the results of the earlier work of Euclid.

We basically are doing similar things to meridian theory, reformulating ancient meridian theory in the language and concepts of quantum theory. We attempt to provide a logical framework to deduce practical results from meridian theory, which is still being used by Oriental doctors to treat diseases. Wherever we can, we support the results from research using modern experimental methods. These results include technical medical terminology, but these may be easily understood by the general readership. We present our experimental and clinical findings in table form. General readers may skip these, read the highlights of the tables, and concentrate on the text and figures.

Meridians and quantum: a dual analysis with taichi diagrams

The taichi diagram was invented about one thousand years ago. It was proclaimed to be the most sophisticated concept in Chinese philosophy. We use them in some of the illustrations in this book.

Every taichi diagram can be subdivided into two parts. Intrinsic to each diagram are many meanings. Let us take Fig 1.1.5 as an example. The convention is that the upper half of the taichi is yang, and the lower half is yin. In this book, meridian theory is the dominant theory and is the yang. Quantum theory is the underlying theory and is the yin. Inside yang there is always a bit of yin (not drawn). Inside meridian theory there is quantum theory. We attempt to explain meridian theory with quantum theory. Inside yin there is a bit of yang (also not drawn). And so inside quantum theory is a bit of meridian theory. Within the formalism of quantum theory there are many aspects that comprise the function of meridians in a living body.

Meridian theory comes from clinical experiences accumulated from the past. Quantum theory is new. Its verification comes from scientific experiments.

Within the old there is something new. It is expected that using quantum theory to explain meridians will greatly enrich the content of quantum theory. Within the new there is something old. Although quantum theory is an invention of the twentieth century, the concept of duality in quantum theory dates back thousands of years. Scientific experiments that are summarized in this book have roots in the clinical experiences of the past. Within the Canon of Medicine, where all the clinical experiments are codified, there are seeds for new scientific theory based on the principle of quantum theory.

The general philosophy associated with taichi and wuchi is summarized in Chapter 5. We give one analysis for one example. Readers are welcome to make such dual analyses for each taichi diagram in this book. They might find new meanings hidden within the diagrams, which may not have been discovered before.

The molecular structure of meridians

Q1.2.1. What are meridians? Are they artificial, like the meridians one uses on the maps of the earth? Or are they real, like the nervous system or blood circulatory system, as ancient Chinese medical literature assumes?

At the moment this is the greatest puzzle concerning the nature of acupuncture. We know from anatomy that there is a blood circulatory system because we find there are veins and artery-like tubes in which blood flows. From anatomy we find no sort of tubes at the purported site of meridians, where some substance called qi is supposed to flow. We learn from anatomy that there is a nervous system, because there are specialized cells, which are long and narrow, called nerve cells, comprising the nervous system. But we do not find any other kinds of cells than those found in normal tissues at the site of meridians.

Q1.2.2. We know that the constituents of meridians must be similar to those in the tissues around them. But do these constituents have some special properties that the tissues around them do not have?

Yes, they are different. They react differently to light propagation. They transmit electrical signals differently and have different mechanical properties. Let us go back to traditional Chinese medical theory to find how different they are. In traditional Chinese philosophy of medicine, the most crucial concept is yin and yang. How does the concept of yin and yang apply to the constituents of the meridians? One of the simplest and most logical hypotheses is to identify yin with a negative charge and yang with a positive charge. So our hypothesis is that constituents in the meridians are electrically polarized into negative and positive charges.

The polarized constituents could be tissues, cells, interstitial fluid, clusters of molecules, or molecules. It has been suggested that they could even be something like liquid crystal (Fei, 1998). As long as meridians are made of some kind of polarizable medium, most of the content of this book remains valid. We shall assume a polarizable medium for our application of quantum theory.

Q1.2.3. Since most of the body is made up of water, could this polarized medium be water?

Yes, it could be water. The water in the meridians may have special electrical properties that are different from ordinary water. Such water has been found, and it contains stable water clusters. These stable water clusters are clusters of water molecules that are polarized, i.e., negative and positive charges are concentrated on separate poles of a molecule. (See References, bibliography for further reading, for a detailed description of the properties of and evidence for the existence of stable water clusters.) If we assume these stable water clusters are the main constituents of meridians, we can easily explain the special properties of meridians (see Figure 1.2.2 and Table 1.2.0). If it is water, it is more transparent to light than the tissue around it. It is electrically charged, so it will respond differently to different electrical signals. Oscillations along the meridians, which are identified with qi in traditional Chinese medicine, are like waves in a river. They can be stopped by mechanical pressure.

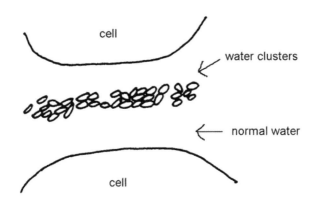

Fig. 1.2.2. Possible model of meridians running between cells.

Q1.2.4. Can you describe those optical properties of meridians that are different from surrounding tissues?

Meridians are much more transparent to infrared light. Researchers at Fudan University in Shanghai have done an experiment with a specimen from a lower limb of a human body (Fei 1998). They found that along the fiber axis of the stomach meridian, the transparency of infrared light is 62 percent, whereas perpendicular to the fiber axis the transparency is only 0.4 percent. There is a more than 150 times difference in transparency whether infrared light travels along the meridian or perpendicular to it

A group led by Zhang D of the Institute of Acupuncture and Moxibustion in Beijing took infrared images of the backs of normal healthy persons. They found that 57 percent of the volunteers showed central lines at the back where the DU meridian runs (Zhang 1996). (See Table 1.2.1 for a summary of related works.)

Since this kind of infrared light can only penetrate normal skin by 1 mm, the transparency of meridians suggests that there is a water solution in the meridians that is transparent to infrared light.

Highlights of tables in text

We have searched extensively for clinical tests and scientific experiments done on animals in the last fifteen years in many different countries. They are done in the tradition of orthodox Western medicine and science and use modern scientific equipment, like ECG, Magnetic Resonance Imaging and immunoreactivity. Many of these are pioneer studies and can easily be improved upon.

Most of the articles about the studies are written in technical terms for the expert and are not easily understood by the layman. Nevertheless their results form the modern scientific foundation for the field of acupuncture and Oriental medicine. So we have included in the text some highlights of the results. In Appendix A you will find tables with details of the tests and experiments.

Highlights of Table 1.2.1. Meridians as optical fiber – optical properties of the meridians

- Some meridians are weakly luminescent. Some meridians of some people are visible using infrared imaging. One experiment finds that a meridian is transparent to infrared radiation in 2.66 um and 10 um, but opaque in its transverse direction.
- Water and normal tissue are opaque to infrared radiation at 10um. Our special water clusters act like a solid, or liquid crystal (see References), and could be transparent.

Q1.2.5. What are the electrical properties of meridians that are different from their surrounding tissues?

Meridians have low impedance. On normal skin, the impedance is lower at acupoints than on neighboring points. Normal skin may have an impedance of 600 kilo-ohm, and at an acupoint the impedance may never exceed 100 kilo ohm. It has been found repeatedly in Japan, Germany, Korea and China that most of these low impedance points lie along the meridians or within 5 mm from the meridians. Recently these low impedance lines were also found in animals, such as sheep, pigs, cats and goats. (Table 1.2.2.)

It is even more interesting to study the effect of electrical signals transmitted through the needle in acupuncture. If one uses ac voltage of a few volts on the needles, prominent effects come from signals in frequencies from 1 Hz to 100 Hz. Currently acupuncturists seldom use frequencies much above 100 Hz. A normal human impulse might be to use a higher frequency, with the idea that the higher the frequency the better, as in the case of transmission of radio signals, where one goes from kilo Hz in the long wave band to mega Hz in the short wave band. In the transmission of TV signals, the frequency goes even higher. However for acupuncture to have an effect, one stays in the low frequencies. Furthermore, the effect of acupuncture depends strongly on the frequency of the AC signals. A 100 Hz signal may have the opposite effect from a 2 Hz signal. (See Table 1.2.3 and 1.2.3A for details.) This may suggest the existence of resonance of constituents in the meridians at these low frequencies.

This could be explained by the fact that different water clusters have different resonance frequencies. Only the right water clusters at the right acupoint will emit the wave with the right frequency to excite a distant organ, which has the same kind of water clusters with the same resonance frequency, to function in a specific way. The wave travels through the meridians.

Highlights of Table 1.2.2. Meridians as conductors – evidence for low impedance along the meridians

- Low skin impedance was observed along three meridians and ren and du meridians.
- Low impedance lines and high percussion sound lines coincide on meridians.
- Acupuncture at acupoint PC3 on the pericardium meridian lowers impedance of the pericardium meridian from 52Kohm to 9Kohm
- Low impedance along meridians has been measured by computerized methods and is highly repeatable.
- Water or aqueous solutions conduct electricity better than normal skin tissue.

Highlights of Tables 1.2.3, 1.2.3A. Summary: difference between 2 Hz and 100 Hz in electro-acupuncture.

Meridians act like cable in a cable television system. Different frequencies in the transmission line will give different programs in the receiving television set, as different frequencies in meridians will have different effects on the related organs or physical systems.

- 2 Hz increases somatostantin (SOM), decreases calcitonin gene-related peptide (CGRP); 100 Hz inhibits SOM, and has no effect on CGRP.
- 2 Hz is good as an analgesia for a rat, but not 100 Hz.
- Different frequencies produce different types of opioid. 2 Hz is mediated by mu- and delta-receptor; 100 Hz is mediated by kappa-receptor.
- The hypothalamus plays a role in mediating low, but not high frequency in electro-acupuncture (EA) analgesia.

Q1.2.6. How are meridians different from their surrounding tissues in mechanical properties?

If you jam your finger in a doorway, your nervous system will carry a signal from your finger to your brain. Your do not feel anything traveling along your arm. People do not feel electrical signals that move through the nerves. However it is a quite common reaction of patients to feel sensation along meridians when a needle is inserted into the acupoints. So this propagation of sensation is unlikely to be electrical signals moving via the nervous system.

Furthermore, we can block the propagation of sensation by pressing mechanically on the path of a meridian. The effect of propagation of sensations has been measured with modern equipment, such as the electroretinogram. (See Table 1.2.4 for details.) So this propagation of sensation must be a mechanical property of the meridians. This suggests the oscillation or movement of some fluid, flowing along the meridians like waves flowing down a river.

Highlights of Table 1.2.4. Meridians as plastic tubes with water: propagation of sensation along meridians (PSM)

When a needle was inserted into the acupoint, some patients would sometimes feel a sensation radiating from the acupoint. The sensation did not propagate randomly in any direction. It propagated along the meridian where the acupoint is.

- Youngsters feel PSM more distinctly and more frequently than adults.
- The improvement of vision from acupuncture among youngsters who feel PSM is much better than among youngsters who do not feel PSM.
- Acupuncture on the hand at acupoint Hegu LI4 of the large intestine meridian stimulates the retina, as was recorded by an electroretinogram (ERG).
- Mechanical pressure: if mechanical pressure is applied on the acupoint between the hand and the eyes along the large intestine meridian, the signal is blocked, as measured by ERG.

These studies seem to suggest that the qi flows in the meridian similarly to how a water wave might flow in a plastic tube. The water wave can be stopped by squeezing the plastic tube in the middle. Since anatomically there have been no tube-like objects found, meridians may be more like underground water, with no definite boundaries.

Q1.2.7. Do meridians behave differently towards ions as well?

Yes they do. The negative and positive charges in a polarized medium are separated. So the negative charge will tend to attract external positive ions, whereas a neutral medium, which is not polarized, will not be able to retain external charges as easily. It has been found that meridians indeed have such properties. They tend to attract hydrogen ions and calcium ions. (See Fig. 1.2.5 for illustration and Table 1.2.5 for details.)

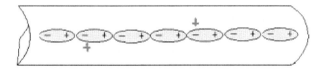

a. Hydrogen atoms attach predominantly to the meridians

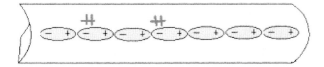

b. Calcium atoms predominantly diffuse along the meridians.

Fig. 1.2.5. Positive ions are found to be attached more along the meridian than to tissues surrounding it.

Highlights of Table 1.2.5. Constituents inside meridians are charged – evidence for meridians, ionic and others

- Calcium ions are found to concentrate along the stomach and gallbladder meridians, by the method of external proton beam induced X-ray emission (PIXE).
- Concentration of hydrogen ions is found to increase along the heart and pericardium meridians for rabbits suffering from arrhythmia induced by aconitine.
- The mast cells were found to concentrate more along the meridian lines in 19 amputated limbs of patients and rats.

A possible explanation is that polarized water clusters in the meridians, with positive and negative charges at their ends, can easily attract and hence concentrate the calcium and hydrogen ions.

Q1.2.8. Does sound also propagate differently in meridians?

Yes. Prof. Zhu Zong-xiang of the Chinese Academy of Science and Prof. Hao Jin-Kai of the Yanan Institute of Acupuncture have demonstrated that sound propagates differently along meridians. They tap one point of the meridian and hear the sound, with a microphone, at another point of the same meridian. The sound is different if one taps points on the meridian versus points not on the meridian. They actually use this difference to determine the positions of meridians, and they find that they are the same as described in ancient classic literature as well as those found by low impedance methods. (See Acupuncture Meridian Biophysics – Scientific Verification of the First Great Invention of China, Beijing Press 1989.)

Q1.2.9. How do all these different properties of meridians fit together?

The existence in meridians of stable water clusters which have a permanent dipole moment (Fig. 1.2.6) are easily consistent with the above-mentioned properties. Water conducts electricity better than tissues. Electric dipoles attract ions better. The most interesting property of stable water clusters is that they have a range of resonance frequency from 0.1 Hz to 100 Hz. (See References, Chapter 1, for collection of related papers.) This is just the range required for electro-acupuncture: so far, the frequency range used by acupuncturists in electro-acupuncture is in the range of 1 Hz to 100 Hz. For the very low frequency of one tenth of a hertz, we see such a strong resonance effect only in the measurement of the electrical properties of a qigong master. (See Chapter 4 for more discussion.)

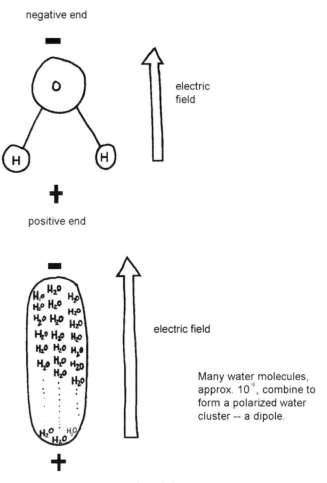

Fig. 1.2.6.

Acupoints and acupuncture

Q1.3.1. What is an acupoint?

In the Chinese language, "acupoint" literally means "hole." Ancient Chinese medical theory uses holes as a simple representation of acupoints. This simple picture is supported by modern laboratory findings in the in the last ten years.

We use the hole on the surface of the skin to peek inside and to make a connection with the body inside. When we use a needle in acupuncture it is essential to insert it into the right hole so that it directly stimulates the meridians and makes them oscillate and carry the signal inward.

Q 1.3.2. What is the relation between acupoints and meridians?

The relation between an acupoint and its meridian is like a well and its underground waterway. A well is a hole in the ground where we can gain access to the waterway underneath. Water underground flows along definite pathways to neighboring regions in the ground. This is similar to how water solutions with water clusters distribute along meridians inside the body.

Q1.3.3. How does this simple picture help us understand the properties of acupoints and meridians?

From this simplified picture, we see how we may influence the interior of the body via acupoints. Substances flow more easily to the interior body through acupoints, as will electrical currents. Indeed, the electrical resistance at the acupoint has been found to be lower than its surroundings. One can verify this fact using devices that can be purchased. These devices can locate the low electrical resistance points on the skin surface and identify them as the acupoints.

In Table 1.3.1, we tabulate the various properties of acupoints. We quote a study done in Taiwan, where researchers mapped the topography of low skin resistance points on rats. The resistance is about 150 to 170 kilo-ohm, which is much lower than the value of 450 kilo-ohm at other points on the skin. So acupoints exist not only on human skin, but also on animal skin.

Researchers in Kaoshing, Taiwan, injected a solution with radionuclide Tc-99m into an acupoint. By following the radionuclides, they traced the path of the solution as it flowed inside the body and found that at acupoint KI-3 the absorption of the solution was much better. An acupoint is like a sink where the radionuclides can flow through much more easily.

A group of Japanese research workers using a scanning and transmission electron microscope found that during acupuncture or moxibustion a large number of immunocytes infiltrate into the area around acupoints.

In experimenting with rats, it was also found that electro-acupuncture influenced the movement of mast cells. The numbers of mast cells in the acupoints became higher than those in area away from acupoints.

There is, then, clear evidence that matter, such as charges, radionuclides, immunocytes and mast cells, can flow in and out of the acupoint area more easily than in other parts of the skin.

The structure of acupoints was examined by a group of research workers at Fudan University in Shanghai using magnetic resonance imaging, X-ray, CT scan, and microscopes. They found that an acupoint was a complex system comprised mainly of connective tissue interwoven with blood capillaries, nerves, lymph vessels, etc. Elements of Ca, P, K, Fe, Zn, and Mn were found concentrated in the deep connective tissue structures in locations corresponding to acupoints.

Highlights of Table 1.3.1. Properties of acupoints – acupoint as a water well, or as a local organizing center

- Size of acupoint as measured by increase of skin conductivity is about 450 microns for humans and 350 microns for rats.
- Topography of acupoints of 6 rats was measured by conductivity. Fourteen Ren acupoints and 17 Du acupoints had resistance in the range of 150-170kohm, which is about three times smaller than that on most other points on the skin, which have resistance of 420K ohm.
- Absorption of radioisotopes via acupoint KI3 on the kidney meridian was much better than that at non-acupoints.
- Magnetic resonance imaging method revealed that an acupoint is a complex system of connective tissue interwoven with blood capillaries, nerves and lymph vessels.

Q1.3.4. What is the purpose of acupuncture? How many different kinds of acupuncture are there?

The purpose of acupuncture is to stimulate acupoints, using different methods, to cause different kinds of oscillations in the meridians. Meridians carry signals embedded in these oscillations to different parts of the body.

Traditionally, acupuncture and moxibustion have been the two most common methods for stimulating acupoints. Acupuncture is done with a needle, which mechanically stimulates the tissue in the acupoints. We expect that an external insertion of a foreign object, in this case, a needle, into an acupoint will cause oscillations in the meridians. Moxibustion gives out heat, which is a form of electromagnetic radiation, given that heat radiation is made up of photons. This heat causes oscillations. An intense beam of photons incident on an acupoint will also cause oscillations in the meridians.

Photons emitted by lasers can also be used for treatment. A patient treated with lasers, however, will feel no heat. For some patients this is a good alternative. The photon beam emitted by a laser has less total energy than the moxibustion heat source, so it should have a smaller therapeutic effect. The difference is that photons from a laser beam have only one single wavelength, whereas photons from a heat source have wavelengths distributed over a wide range of values, which gives less control to the practitioner. A typical laser in laser acupuncture generally has a wavelength of about 800 nm. So far, laser acupuncture has produced quite good results, as listed in Table 1.3.2.

Highlights of Table 1.3.2. Effect of laser acupuncture: photonic stimulation of meridians

- Reduces post-operative vomiting with laser acupuncture at PC6.
- The activities of the cellular immunity and humoral immunity of rats were enhanced by He-Ne laser by using a transverse electron microscope (TEM).
- 70% of patients with initial manifestations of cerebral blood supply insufficiency derived benefit from laser acupuncture.

Laser, a new darling in medical science, is now used more and more frequently as an acupuncture device. It is used as a new of kind of photonic stimulation at the acupoint, without insertion of a needle, with positive effects. According to our theory, it is important to control the frequency of the laser pulse, the time duration of the pulse and the power of each pulse. Different frequencies may have different effects, as in electro-acupuncture. The duration, power and timing of the pulses may affect the excess and deficiency of the particular meridians treated. [End of table highlights.]

If the purpose of the acupuncture needle is to cause oscillations, and oscillations are made by molecules whose charges are polarized, logic suggests that applying external ac electrical signals would also cause oscillations and achieve similar results as acupuncture. In fact this has been done, in at least in two ways. One way is called transcutaneous electric nerve stimulation (TENS), where the electrodes are placed on the surface of the skin. The effect of TENS is tabulated in Table 1.3.3. The other way is the application of an electric potential directly to the needle, which is called electro-acupuncture (EA). For treating animals, acupuncture is usually done with an electric potential, or EA.

Highlights of Table 1.3.3. Effect of transcutaneous electric nerve stimulation (TENS) at acupoints

- Positive effect on therapy-resistant hypertension: mean diastolic blood pressure and systolic blood pressure decreased significantly.
- The opioid requirement for 100 women after hysterectomy or myomect-omy was decreased by 35% and 38%.
- TENS with surface electrodes significantly increases pain thresholds of skin and fascia but not those of muscle or periosteum.

- High frequency 100 Hz but not low frequency 2Hz was effective in ameliorating muscle spasticity.

Evidence shows that application of electric current by electrodes on the surface of acupoints, as in TENS, has similar effects to electro-acupuncture, where needles are inserted into acupoints. This is understandable, because acupoints are like water wells that conduct electricity from the surface of the acupoints to the inside of the acupoints, then to meridians. There would be only a smaller effect if applied on the surface because of the additional resistance between the surface and the inside structure of the acupoints.

In many studies it has been found that EA is better than TENS, and both EA and TENS are better than manual handling of the needle.

Many patients are afraid of needles and do not like to see them piercing their skin. Then acupressure may be an alternative. Pressure is applied on the acupoints by pressing hard with fingers. This gives mechanical stimulation to the acupoint region. But since it is not inside the skin, it is further away from the meridian. So the effect is generally not expected to be as good. However, many studies verify that acupressure is an effective treatment. This is tabulated in Table 1.3.4.

Highlights of Table 1.3.4. Evidence of acupressure – mechanical stimulation of acupoints

- Thumb pressure on PC6 has an anti-gagging effect on dental patients.
- 71% of women reported less intensive morning sickness and shorter duration of symptoms by using wristbands.
- Acupressure on PC6 reduced nausea and vomiting during pregnancy for 60 women.
- Pressure on acupoints decreased systolic arterial pressure, diastolic arterial pressure, mean arterial pressure, heart rate and skin blood flow on 24 healthy male volunteers.
- Significant difference was found in regard to nausea experience and nausea intensity for women undergoing chemotherapy, in favor of acupressure.
- Significant reduction in the frequencies of nocturnal awakening and night wakeful time for 84 institutionalized residents.

Q1.3.5. Acupressure is less effective than acupuncture, but it has the advantages of not involving the insertion of a needle and can be performed by patients themselves. Needles for an acupuncture treatment generally stay inside the skin for about thirty minutes, and are then taken out. Is it possible to use a much longer period of stimulation than thirty minutes?

Yes, this is possible if we inject a foreign substance, which may be liquid or solid, into the acupoint area, which will not be taken out.

The foreign substance serves as a stimulant, either causing muscles to contract in order to eject it, or inducing biochemical reactions. These cause oscillations in the meridian. The stimulation stops only when the foreign substance is absorbed and has gone to an area away from the acupoint. If we used a solid substance like catgut, the stimulation wouldn't stop until the catgut was totally absorbed by the body. If a liquid injection were used, the liquid could be benign. It could be some medicine, or it could even be a poison, like bee venom. Then there would be an added medicinal effect, which couldn't be easily separated from the purely mechanical stimulation. These are also effective and are tabulated in Table 1.3.5.

Highlights of Table 1.3.5. Evidence of injection at acupoints: stimulation at acupoint with foreign substances

- Injection of bee venom into acupoint ST36 on the stomach meridian of rats has anti-inflammatory and antinociceptive effects, promising therapy for long-term treatment of rheumatoid arthritis.
- Pain disappeared for patients with biliary colic, with water injection in LI14, GB24 and RN14.
- Warm needling and point-injection with Zhuifengsu enhanced the immunological function of patients with rheumatoid arthritis by increasing activity of natural killer cells and IL-2 value.
- Using acupuncture on endermic points of the head with extract of positive allergens had significant curative effects on 105 cases of allergic rhinitis.

Moxibustion

Moxibustion is as ancient as acupuncture. Its efficacy has now been confirmed by modern clinical tests (tabulated in Table 1.3.6). Moxibustion can improve the immune system, induce analgesia, regulate renal function, cure chronic colitis, protect against liver injury, and even induce gene expression.

Highlights of Table 1.3.6. Evidence of Moxibustion – meridians as infrared-transparent optical fiber

- On immune system: Moxibustion at acupoints Qihai RN6 and Tianshu ST 25 inhibited the expression of IL-1 (beta) and IL-a6 m RNA in experimental ulcerative colitis rats.
- Moxibustion at acupoints Guanyuan RN4 on Sarcoma S180 ascitic mice increases the decreased erythrocytic C3b receptor rosette-forming rate, decreases the raised immunocomplex rosette-forming rate, and increases activity of erythrocytic immunosuppressive factor in the tumor-bearing mice. Hence moxibustion strengthens erythrocytic immunity.
- On tumor bearing mice there was an instant elevation of serum ACTH and beta-EP from moxibustion at Guanyuan RN4.
- Moxibustion at acupoint Guanyuan RN4 of tumor bearing mice promotes hyperplasia of the pituitary and adrenal glands, stimulates the secretion of beta END from the pituitary and adrenal gland, and increases the level of

serum beta-END significantly.
- For arthritic rats, moxibustion at acupoint Shenshu BL23 could lighten a local inflammatory reaction, eliminate swelling, prevent or reduce the polyarthritises, maintain the weight and shorten the course of the disease. It could recover and promote the effects of concanavalin-inducing splenic lymphocyte proliferation in rats. It could also promote interleukin2 production and decrease IL-1 contents.

On analgesia (Table 1.3.6, cont.)

(a) Moxibustion-induced analgesia was studied in rats that were urethane-anesthetized. Single unit extra-cellular recording from neurons in the trigeminal nucleus caudalis were obtained from a micropipette. Suppression was observed on both a wide dynamic range and nociceptive specific, but not on low-threshold mechanoreceptive units. Moxibustion induced moderate suppression with a long induction time. It suggested that noxious inhibitory controls may be involved in the analgesic mechanism.

(b) The analgesic effect of moxibustion was measured by the latency of tail flick threshold (LTH) of rats. When the temperature of the surface was modulated within $38\text{-}39^\circ$ C and $43\text{-}44^\circ$ C, LTH increased $17.7 +/- 2.1\%$ and $22.2 +/-2.5\%$ respectively, after 5 minutes ($p<0.05$).

On renal function, colitis, ulcer, neurons and gene expression

(a) The effect of moxibustion at acupoints BL15 and BL27 was studied on spontaneously hypertensive rats. Urinary volume was increased for BL15, but decreased for BL27. Urinary secretion of Na+ was decreased for BL15 and BL27. Systolic blood pressure was decreased for BL 15, but not for BL27. Plasma levels of aldosterone and renin activity were increased, and atrial natriuretic peptide was decreased for BL15. Plasma levels of aldosterone and atrial natriuretic peptide were increased for BL27.

(b) The effect of moxibustion at acupoint RN4 on the function of MDR gene product P-glycoprotein P-170 of mice with S-180R adriamycin-resistant tumor cells was studied. A weak inhibition was found when moxibustion was performed at RN4 alone, and a very significant inhibition was observed in the presence of low dosage of verapamil but not at high dosage. [End of Table 1.3.6 highlights]

All the types of stimulation can be classified into two broad categories: with, or without foreign substance injected into or buried in the skin. Stimulations at the acupoints can be classified into mechanical stimulations or non-mechanical stimulations, which are mostly electromagnetic. Mechanical stimulations can be done inside the skin or on the skin surface. Needles are used inside the skin. Acupressure is applied on the skin. Electromagnetic stimulations may be done with or without photons. Photons, as discussed above, may come from lasers or from moxibustion. Electromagnetic stimulations include electric, magnetic and electromagnetic stimulations.

Electrical force is generally much stronger than magnetic force inside the human body. A magnetic field can influence the body through its interaction with magnetic materials or electric current inside the body. Since there are few magnetic materials inside the human body, and whatever electric currents there are inside a human are very small (much smaller than one milliamp), the magnetic effect would be small. But sometimes a small effect in the right place and at the right time may be quite effective. Nevertheless, in most cases, an electric effect should be much stronger. So electric rather than magnetic stimulations are currently the preferred way to treat patients.

Q1.3.6. So far we discuss acupoints in a mature human body. How do they develop from a single fertilized egg in an embryo? Or in a much longer time scale, how do they arise in the evolution of life?

Dr. Charles Shang developed a theory about how acupoints are developed in embryo. As an embryo develops, it retraces all the steps in the evolution of its species. Shang proposes that as a single cell evolves into a complex structure with many cells, a small group of cells will act as the local organizing center that controls the surrounding larger group of cells. Local organizing centers are known to have low electrical resistance and a high concentration of gap junctions. Gap junctions are structures between cells that chiefly perform a communication function among cells. Local organizing centers evolve into acupoints as the embryo grows. Naturally, acupoints would have low electrical resistance and be able to transmit signals, as we discussed earlier in this section. Acupoints are also converging centers of electric current and magnetic flux.

According to the "gradient model" in developmental biology, organizing centers are found in extreme points of curvature on the body surface. Almost all the extreme points of the body surface curvature are acupuncture points. There are two kinds – convex, or concave, depending on whether the surface sticks out or caves in.

Examples:

The acupoints EX-UE 11 Shixuan, Ex-LE Qiduan, ST 17 Ruzhong, ST 42 Chongyang, ST45 Lidui, SP 1 Yin Bai, SP 10 Xuehai, Du 25 Suliao, and EX-HN3 Yintang are convex points.

The acupoints RN17 Danzhong, KI 1 Yongquan, LI5 Yangxi, LU5 Chize, LU 7 Lieque, LU8 Jingqu, LU10 Yuji, SI19 Tinggong, TE 21 Ermen, GB 20 Fengchi, GB30 Huantiao, BL 40 Weizhong, HT 1 Jiquan, SI 18 Quanliao, BL 1 Jingming, RN 8 Shenque, and ST 35 Dubi are concave points.

As the embryo develops, organizing centers evolve into a network, which becomes the growth control system. The formation and maintenance of all the physiological systems are directly dependent on the activity of the growth control system. The growth control system evolves into the meridian system. The evolutionary origin of the growth control system is likely to have preceded all other physiological systems.

Meridians are cellular networks, which are connected together to regulate growth and physiological function. They lie at the boundaries between different muscles, as discussed in ancient Chinese medical literature. They are folds on the surface of various domains of the body, or boundaries between structures.

- Part of the lung meridian runs along the borders of the biceps and brachioradialis muscles.
- Part of the pericardium meridians runs between the palmaris longus and flexor carpi radialis muscles.
- Part of the gallbladder meridian runs between the sternocleidomastoid and trapezius muscles.

Dr. Shang called his theory morphogenetic singularity theory. It contains many exciting features that incorporate modern developments in fields such as morphogenesis, embryogenesis and ontogeny, some of which we have discussed above. Readers who are interested in more detail should consult the references for this chapter.

Q1.3.7. How does the idea of yin and yang help us understand acupoints and meridians?

In this book we use taichi diagrams extensively as tools. This is in line with the philosophy of traditional Chinese medicine, where yin and yang are the most fundamental concepts. Yin and yang were first proposed in the Book of Change (the I Ching) more than two thousand years ago. They were organically combined into the taichi diagram about one thousand years ago. The taichi diagram is one of the most profound philosophical concepts invented. In a taichi diagram, yin and yang are separated by a wavy line, indicating that the separation of yin and yang is not rigid. Yin and yang interact with each other continuously. Inside yin there is yang, and inside yang there is yin. Relations among phenomena and concepts are greatly clarified when placed in the yin and yang of a taichi.

In the following discussion we shall clarify the specific meanings of a few specific figures in terms of the general properties of taichi diagrams. We do not have space to explain all of the figures. Readers are encouraged to read a deeper meaning into each taichi presented in this book. In the end, I hope readers will be able to apply taichi diagrams to practical problems they encounter in their daily life.

Let us now expound upon the meaning of Fig. 1.3.7, where we set the acupoint as yang, and the meridian as yin, within a taichi. When we put a needle into the patient, the acupoint is the active aspect. The attention is on acupoints, the yang. The meridian is lurking in the background, and is hence the yin. The placement of the acupoint is decided by the meridian. The effect of the acupoint relies on the function of the meridian. The visible part is the acupoint; the invisible part is the meridian.

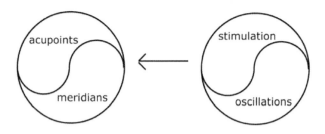

Fig. 1.3.7. When one stimulates acupoints, the oscillations of meridians are affected.

We stimulate the acupoint to induce oscillations in the meridian. Stimulation and oscillation then occupy the yang and yin of a taichi. Within yang there is yin. Within stimulation is the force of additional oscillations in the meridian. Within yin there is yang – the oscillation of the meridian carries the information of oscillation from that particular acupoint.

Quantum theory, predictability and uncertainty in life

Q1.4.1. Why do we need quantum theory in understanding life and acupuncture?

We need quantum theory for the following reasons:

1. Acupuncture uses a needle without putting any foreign substance into the body. Many clinical tests and scientific experiments with animals, nevertheless, have shown that biochemical reactions occur inside the body when electro-acupuncture is used. Electro-acupuncture only moves electrons and ions, which results in biochemical reactions. Explaining biochemical reactions through the movement of electrons and ions belongs to the domain of quantum theory. So quantum theory is needed to explain the mechanism of acupuncture.

2. When we investigate the finer effects of oscillations in any classical waves, as in light or sound, we inevitably come to quantum theory. Sound becomes phonons, and light becomes photons. So when we study the finer effects of the circulation of qi in meridians, it is inevitable that we come to the smallest lump of energy of qi, which we call qion. Since qion is defined in exact terms of physics, experimental equipment could easily be designed to measure its many effects in the not too distant future.

3. Human activities are typically described as having distinct states. A person may be in the state of sorrow, of anger, of sleeping or waking, in a state of deep concentration, etc. Such distinct states are described naturally as discrete quantum states in the quantum theory of a system.

4. Future events in a human life are best described by probability, as intrinsic in quantum theory, and not by certainty, as in classical theory.

There are measurements of infrared radiation emitted from the hands of qigong masters that does not obey black-body* radiation laws. This could easily be explained by the radiation having a component of coherent infrared radiation, because it is emitted through the acupoints, from ordered water clusters. Such a quantum effect should be confirmed or denied in future experiments. (*A black body is, in physics terminology, one that emits the maximum amount of heat radiation.)

Q1.4.2. Can quantum theory predict the future?

It can show us various probabilities only. Let us take, for example, the case of the future of a young person. We raised our third child in Los Angeles. Our neighbor has a 10-year old daughter. Since Hollywood is in Los Angeles, children in Los Angeles are more influenced by movies than elsewhere. This child's hope is to become an actress when she grows up. Her future will be shaped by three elements:

1. The possibilities: It is possible for her to become an actress. It is also possible for her to become a nurse, a computer programmer, or a professor. Her future is not certain. There is a probability associated with each possible future.

2. The goal: She set the goal of becoming an actress.

3. The means: There are various ways to reach her goal. When this girl grows up, she may go to college, or she may go to work for a movie studio.

In quantum theory there are three basic elements in all physical processes. They correspond to the three elements of this little girl's process of growing up:

1. Given the same initial state, there are various possible final states. It is not predetermined which final state a given initial state will evolve into. For each possible final state there is associated a probability that can be calculated. The initial state for this girl is that she is ten years old. Her final state may be working as an actress. Other final states could be nurse or professor.

2. Probability is meaningful for a physical process where both initial and final states are specified. It is not possible to infer from the initial state of a physical process what the final outcome will be. We cannot tell what the girl would become if she did not set her goal as becoming an actress. Once she sets her goal, then we may guess at what her chances are of becoming an actress.

3. After the initial and final states are specified in a physical process, we need to know the internal interactions of all objects involved in the physical process as well as their external boundary conditions. From the knowledge of internal and external conditions it is possible to use a principle called the Action Principle to calculate a particular solution for this physical process. (See References for Chapter 1 for additional reading.) For the girl to succeed as an actress, we have to know her internal, or personal qualities: her talent and her appearance. Then her external financial conditions, such as her family's income and her parent's support, will determine whether her best course will be to go to college or to go to work in a movie studio.

Q1.4.3. Can you elaborate on certainty and uncertainty in life?

In nature, there are many very certain things. The sun will rise from the east in the morning and set in the west. The movement of the sun in the sky is predetermined by the laws of classical physics. Its trajectory is certain and definite. Human activities are more complicated and are not easily predicted. The laws of quantum theory may be used to describe the uncertain aspect of human activity.

The movement of any part of a human that can be substituted by a mechanical device, like the hand or leg, is described by classical physics. Its motion is certain. The behavior of crucial molecules in the human brain, such as the firing of neurons, is determined by the laws of quantum theory and is not certain, so we can calculate only the probability that a certain event will occur or not occur.

Q1.4.4. How do human beings differ from, say, the sun?

When we say certain processes, like sunrises and sunsets, occur with certainty, we mean that, given an initial state of an object, there is only one unique final state. Initially, in the morning the sun rises in the east; it can only set in the west. It will not set in the north or south. There is only one definite final state that the sun will come to in the evening. For events that are described by quantum theory, like the thinking process of a human being, there are many possible final states for any given initial state. When a person is diagnosed with cancer, for example, he may become miserable, resigned, or unmoved. He may die soon, or he may survive. There are always several possible endings for any given event.

Q1.4.5. Hope, then, is associated with an uncertain future, is it not?

When the future is certain, there is no room for hope. It is certain that the sun rises in the morning and sets in the evening. We cannot hope for it to be otherwise. An individual's future, however, is never certain. Therefore, there is always room for hope. For example, if you catch a fatal illness, there is always a chance that you may recover.

Choice and free will also play a part in giving us hope. A person with lung cancer, for example, can choose to have an operation to remove part of his lung and can hope for recovery. The molecules in the neurons involved in making the decision for the lungs to recover or not can behave in various possible ways. The behavior of the molecules is governed by quantum theory and is indeterminate. So here we have room for free will to play a part, and there is hope that one may recover.

Given the indeterminacy when human choice or other high-level human functions come into play, we may use quantum theory to explain the course of events.

Q1.4.6. Can you us give some examples that show how the language of quantum theory is quite consistent with daily experience?

It is common for an individual to have pain. He may have a headache, or his back may cause him pain that is unbearable. He can use several methods of relief. The easiest and most common way is to take pain-killing pills. The pain will go away, but only the symptom is relieved. The relief is most likely temporary, and for some people, painkillers have undesirable side effects. Or the individual may seek the help of an acupuncturist who can locate the origin of his pain and cure it completely without any side effects. So, given an initial state of pain, there are various final states of no pain that can evolve. A patient has a choice of which final state he wants to be in.

An overweight person, when presented with a dessert after a hearty meal, always has the choice of saying no. If he goes on a weight reduction program, he may come to a variety of final states in one month's time. His weight reduction may be temporary or permanent. If his program includes the use of a drug, he may experience side effects. He has a choice of different ways of reducing his weight.

When we get old, we may wish to live longer. But a longer life is not a unique state of affairs. There are many states of longevity. We could have a long life with vigorous physical and sound mental health. We could have a long life with a physical handicap and forgetful mind. We could have a long working life, or we could have a completely relaxing life without having to work. There are many final states to choose from as to how to spend old age. In Chapter Four we shall discuss the possibility of attaining vibrant health, even into old age.

Q1.4.7. Can you name one feature of quantum theory that is not obvious, but that affects our actions every day?

One feature that is not obvious is that as we move, we may be aware of infinite past, infinite future, and all our surroundings. This is inherent in Feynman's space-time approach to quantum theory. As a particle moves an infinitely small step from A to B, it travels at infinite speed to every part of the universe, from infinite time in the past to infinite time in the future. In other words, when the particle moves from A to B, it goes through all possible paths in between (Fig 1.4.7).

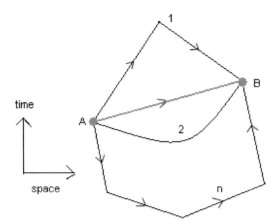

Fig. 1.4.7. In Feynman's space-time approach to quantum theory, for a particle to go from A to B, it goes through all possible paths in space-time, path 1, path 2......path n......

The probability that the particle actually goes from A to B is the summation of all probability amplitudes of going through all possible paths. This movement in an infinite number of paths is not real. It is virtual. Nevertheless, if there is a boundary that prevents the particle from entering, it knows where the boundary is. All the paths that the particle travels are virtual paths. They do not violate relativity, because there is no information being transmitted faster than the speed of light.

Take the example of an excited neon atom in the middle of a He-Ne laser cavity. When the excited atom decays, it emits a photon. The wave function of the photon has a normalization volume, which is equal to the volume of the laser cavity. Since the neon atom is far away from the sides of the laser cavity, classically it is impossible for it to know where the boundary of the laser cavity is. However, in quantum theory, this does not have to be the case. As soon as a photon is emitted by the excited neon atom in the middle of the cavity, it will travel instantly, at infinite speed, in all directions, and come back. So it is possible for the photon to know its boundary, and its volume normalization can be equal to the volume of the laser cavity.

Let us extend this idea and apply it to daily action. If, for instance, we drive to visit a friend, at a traffic intersection we have to decide whether to turn left or right. We don't make the decision based just on seeing the traffic sign. We're also aware of where we're going and where our friend's home is. This has actually been accounted for by our wave function, which has already traveled to the limit of the boundary of where we can go and has come back, giving us additional sensory information. We may not be aware of this, but our decision has been informed by this data about the boundaries of our targeted surroundings. So we have used data collected by our eyes, our ears, and our wave function.

Q1.4.8. Can you say more about the human wave function?

Our wave function is of two types. The first is that of the physical body, such as the movement of our hands and feet. There is a wavelength associated with any object, which is inversely proportional to its mass. Quantum effect is prominent only at the distance comparable to the wavelength. For a heavy object, like a hand, the wavelength is extremely small. So the quantum effect of a hand is very small.

The second type of human wave function comes from the oscillations of the meridians. We shall concentrate on the wave function that describes the oscillations. These resemble, in function, the waves that cause a sound when a piano wire is struck with a piano key, or the sound emitting from vibration of our vocal chords.

Some oscillations in the meridians have no mass. They radiate from the body and can easily propagate through space, like ordinary sound or light. It is this second part of the wave function, generated by the oscillations of meridians, we believe, which may be responsible for the highest level activity of a human being.

Q1.4.9. Can you describe one general property of wave functions that is relevant to understanding human behavior?

There is one very general symmetry property of wave functions. A wave function may be characterized by symmetry or the lack of it. There is the symmetric wave function and the anti-symmetric wave function. The wave function of a photon is a symmetric function. The wave function of an electron is an anti-symmetric one. Particles described by a symmetric wave function can share the same state. Generally speaking, particles having a symmetric wave function like to travel in the same direction. In a laser cavity, when there are already many photons moving in the same direction along the laser cavity, any new photon emitted from an excited atom will most likely go in the same direction. In the absence of many photons surrounding it, a photon from an excited atom may go any direction (Fig. 1.4.9). There is very little chance that it will go along the direction of the laser cavity.

In human activity we find something similar. Take the case of meeting with a group of friends one evening and discussing what you are going to do. If the majority decides to see a particular movie, say, "Hidden Dragon, Crouching Tiger," it is very likely that you will go along with them to see the movie. We may call this the "crowd effect."

The crowd effect may be attributed to the symmetric wave function of a human being. Particles that are described by an anti-symmetric function, on the other hand, cannot share the same state. They obey the exclusion principle. Electrons are such particles. An atom is composed of many electrons.

Suppose in an argon atom there are twenty electrons packed outside its nucleus. Let us assume that all electrons have different states. We will label the different states with the letter n, s, and p, as in "s-state," "p-state," etc. Let us say that these labels correspond to the labeling of seats in a theater.

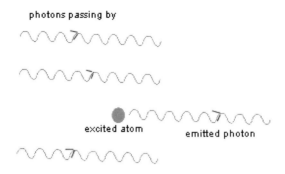

Fig. 1.4.9. Coherent effect: an excited atom in a laser will emit a photon in the same direction as the other photons passing by it.

Each electron has a unique seat in the atomic orbit, so to speak. None of the electrons shares the same state, or the same seat. Similarly in human activity there are many instances where people do not like to share the same state with other people. Using the example of seats in a theater, in the assignment of seats there can be only one seat per person. Each seat has a number and each person who buys a ticket would have a unique number and a unique seat. You cannot assign two persons to a seat.

Such exclusiveness in human activity can ultimately be traced to the anti-symmetric wave function part of a human being.

Conclusion

In quantum theory, certainty and uncertainty exist together. Events may be predicted only as probabilities. Whether an electron passes a certain hole or not cannot be predicted with certainty. But not everything about electrons is uncertain: the electron still carries the same charge. And the conservation of charge is certain to hold.

Chapter 2

Quantum of Life – Life as Vibration

Three levels of existence

Q2.1.1. Who am I?

When I wake up in the morning with sleepy eyes, barely conscious of anything, I probably could not even recall my name right away. But when I wash my face and see my reflection in the mirror, I recognize myself. My face is myself. My appearance is what my friends identify me by, but this part of my identity is only skin deep. My friends know more than my face. My image is only a reminder of what I do, what I think and what I mean to them.

As the French mathematician-philosopher, Rene Descartes, said, "I think, therefore I am." A person's existence involves, at the least, what he thinks. So we can identify at least two levels of existence: a physical, material part, like the body, that one can touch and feel; and a nonphysical, non-material part, the mind, which an observer cannot touch or see, but which we know exists. The physical part is easy to quantify and measure. The nonphysical part is not so easy to quantify and measure.

Q2.1.2. What is the common idea of basic existence in Western culture?

It is common in the West to speak of three levels of existence: body, mind and spirit. The physical level is the most obvious and the most commonly acknowledged in our daily life. Our face is our primary mark of identification. Friends recognize us by our face; our passports use pictures of our face. The Federal Bureau of Investigation uses fingerprints as identification. Most of the time the outer physical appearance is enough to identify us. And now, the inner physiology may be identified by our molecules, such as DNA.

When we discuss the higher-level activities of a human being, however, we need to speak about mind and spirit. These are what distinguish us from animals. Our bodies are not so different from those of animals. It has been shown that we have about 99% common DNA with primates. So there is only 1% difference between our DNA and that of chimpanzees. We believe that our minds, i.e., our thinking and our psychological make-up, are far more complex and much different from those of animals. The distinction between mind and body, however, is vague, and is a source of great philosophical and psychological inquiry.

For many of us who believe that we are measurable by something more than our achievements of mind that can be recorded in books or music, there is something extra, called spirit. The meaning of spirit depends very much on one's belief, rather than on any scientific evidence.

Q2.1.3. In Eastern thinking, are there also three levels of existence?

In the East, qigong masters also speak of three levels of human existence. The lowest level is jin, the second level is qi, and the third level is spirit. Jin means essence of life. It includes all the vital functions of the body, such as sexual energy. By practicing qigong, one can turn jin into qi, and then qi into spirit. So for qigong masters the three levels of existence are very concrete. For people who do not practice qigong or some form of meditation, these three levels of existence – jin, qi and spirit – may be hard to understand.

Q2.1.4. Can you define three levels of existence in terms used in physics, so that they are understandable and accessible to experimental investigation?

Here we propose a three-level model for human beings. The levels will be formulated in analogy with examples from physics, specifically, from quantum theory. Like all models in physics, it is a very simple model. However, there is nothing vague. In principle, everything is measurable and understandable.

We propose, at the simplest level, three states of human existence: the state of a dead body, the ground state, and the excited state of a living being. We'll simplify these into three quantum states. Each state has a specific energy and a well-defined meaning.

Let us illustrate this concept using the example of a hydrogen atom, having three levels of existence. The first level consists of the constituents – the hydrogen atom has a proton and an electron that is unbound to the proton. When the hydrogen atom is dissociated, a proton and an electron are what must remain. The second level of existence is the ground state. In the hydrogen atom, the electron circulates rapidly around the proton when the atom is at its lowest energy state. The third level of existence is when the hydrogen is excited – the excited state. The electron has acquired extra energy, and it circulates in a larger orbit than when it was at the ground state. There are many excited levels of a hydrogen atom, where the electron occupies different orbits around the proton. We group them altogether as the third level of existence.

Q2.1.5. Does a more complicated inanimate object have three levels of existence?

We may describe inanimate objects as having three levels of existence. Take the example of a computer. A personal computer, or PC, has a microprocessor, a monitor, a hard disc and other components. When the parts are assembled into a system sitting on a table, it is a dead PC. All its components can be identified and their functions are well defined. This is the first level of existence of a PC. The second level of existence of a PC is when the power cord is plugged in, and the system is on. But still there is no action. We shall call this the second level of existence for a working PC. It is the lowest energy state, or the ground state. The third level of existence is when the PC is performing work. When you type a manuscript, the PC microprocessor processes information. The monitor shows the words that are being typed. This third level of existence is an excited state of the ground state. More energy is being consumed. Information is being processed at an extremely rapid rate.

Q2.1.6. Can a living object have three levels of existence?

We can identify three levels of existence for a living object, a kidney, for example. If we take a kidney from a dead body, we can count the number of cells it has. All the kidney's components are well known and well defined, except that there is no motion and no life. There is only the physical body of the kidney. This could be described as the first level of existence for a kidney. The second level of existence is when the all the kidney cells are alive, but it is not functioning as a living kidney would, e.g., it is not cleansing blood. This state characterizes a kidney taken from a living person. By itself it is alive, but it is not functioning as a part of a human body. This is the lowest energy state of a living kidney – the second level of existence. The third level of existence for a kidney is when it is inside a healthy person and is functioning properly. It is in an excited state. If we examine a living kidney inside a healthy person, we shall see that the cells are full of activity. Fluids flow in and out of the kidney cells at a rapid rate. The urea is filtered and clean blood flows out of the kidney. The separation of three levels of existence for a kidney is not as distinct as those of the PC. Nevertheless, we shall use the concept of three levels of existence as a good approximation to classify the broad differences between a dead kidney, a barely alive kidney, and an active kidney.

Q2.1.7. Could three levels of existence also be applied to the brain?

The human brain can be seen as having three levels of existence, to greatly simplify. The first level would be a dead brain with many brain cells but no activity. The second level is the lowest energy state of the brain, when it is alive but there is no perceivable action. The brain of a person who is unconscious, or is sleeping without dreams, may be the closest to this state. The third level is when the brain is fully functional. This happens when a person is active, as in talking, walking or thinking. It is an excited state. The brain is fully alive. Obviously there are many excited levels.

Q2.1.8. How about an individual as a whole – does he have three levels of existence?

Let us represent a live individual, as a whole, by the symbol I. His lowest energy state may then be denoted by I_0. P denotes a living individual, whose physical body would remain the same if he were not alive. Here we have three kinds of individua

The splitting of an energy state into many states is a familiar scene in atomic physics. There is fine splitting and hyperfine-splitting of atomic levels, say, of a hydrogen atom. So one may expect that each of the three human states could be split into many finer states.

The excited states of a human can be classified into those that are caused by the oscillation of brain waves and those caused by the oscillation of meridians. This is similar to energy states in molecules. There are rotational states, and there are vibrational states. Oscillations of brain waves may be regarded mainly as forming the human mind. Some of the oscillations of the meridians and their effect may be regarded as the spirit of the human being. So mind and spirit will have a very explicit meaning, which may be subjected to experimentation. They may not be exactly the same as what others refer to as mind and spirit. Defining mind and spirit in this way has the advantage of clarity of thought. These energy states, analogous to energy states of atomic and molecular spectra, are illustrated in Fig. 2.1.8. It is quite clear that further splitting of energy states is possible.

The non-excited and excited states of the brain as mind are generally classified as the subconscious and conscious states. We may use the personal computer (PC) as an analogy. There are parts of the PC that cannot be displayed on the monitor, but their existence is essential for the working of the PC. This corresponds to the subconscious part of the brain.

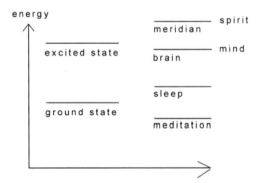

Fig. 2.1.0. Splitting of energy states: The ground state is split into sleep state and meditation state. The excited state is split into excited states from oscillations of the meridians and the brain.

There are parts of a PC that can be displayed on the screen. Everything that this part of the PC does is available for close examination. This would correspond to the conscious state of the brain.

The excited state of meridians is due to energetic oscillations of meridians. We may divide oscillations into two kinds, which correspond to qi and spirit, as discussed in traditional Chinese medicine. Here spirit is used in the sense of "having high spirits," connoting being "very lively or energetic." Spirit in this sense results from oscillations, and hence eventually will be measurable.

an analogy. There are parts of the PC that cannot be displayed on the monitor, but their existence is essential for the working of the PC. This corresponds to the subconscious part of the brain.

There are parts of a PC that can be displayed on the screen. Everything that this part of the PC does is available for close examination. This would correspond to the conscious state of the brain.

The excited state of meridians is due to energetic oscillations of meridians. We may divide oscillations into two kinds, which correspond to qi and spirit, as discussed in traditional Chinese medicine. Here spirit is used in the sense of "having high spirits," connoting being "very lively or energetic." Spirit in this sense results from oscillations, and hence eventually will be measurable.

We use the terms mind and spirit, which are derived from the traditions of East and West, but we use them differently, because we define them with physical concepts, and in principle they will be measurable in the future.

From qi to qions – oscillations as quanta of life

Q2.2.1. What is a fundamental definition of life?

We may define life or its absence by the presence or absence of vital signs. Fundamentally, oscillations are a vital sign. If we have a very sick friend lying on the bed in a hospital, and if we hear him speak, then we know he is still alive. So sound uttering from a person is a vital sign, and sound comes from oscillations of the air.

If the person no longer speaks, however, he may still be alive. If we touch his heart and feel his heart beating, then he is still alive. Heartbeats are oscillations of the heart.

Clinical death is usually defined as the moment the brain stops emitting brain waves. When an electrometer can no longer detect electric signals from the brain, the patient is declared clinically dead. Brain waves are oscillations of electric waves in the brain.

According to classic Chinese medical literature, a person is dead when qi stops. Qi consists of oscillations of meridians. So oscillations are fundamental signs of life, what is fundamentally required for a being or organism to be alive.

Q2.2.2. Meridians are rather complex. Can you use, say, the simple example of a piano wire to illustrate the relationship of oscillations and life?

When you go to a piano shop and ask for a piano wire, the shopkeeper will take out a bag with a curled wire inside. The wire is not ordered, in the sense of being straight. When you take the piano wire home and put it inside the piano, screw it in place, then have a piano tuner tighten it correctly, the wire is then in an ordered state. The wire is the same, but if it is disordered, curled up inside a bag, it won't make a sound when hit. But an ordered wire, stretched tight and straight inside a piano, will make a beautiful sound with a definite frequency when it is hit.

So we may say that a piano string has three levels of existence. When it is lying on a table, it curls and remains motionless. It is in a state of disorder. When the string is strung inside the piano, it is straightened and is in an ordered state. Then when the hammer of a key hits it, the wire oscillates and produces a beautiful sound. Life has come to the wire. There is live music.

The state of disorder is the lowest form of existence. The ordered state in the lowest energy state is the second level of existence. The oscillation mode is an excited state, the third level and highest form of existence.

Q2.2.3. A piano produces sound. Does sound have an independent existence outside the piano?

When the piano wire oscillates, it moves the air back and forth. The sound from the piano wire is carried forward by the oscillation of the air in all directions. The sound has a fixed frequency. For a C note, the oscillation in the air is 254 times a second. When the sound wave reaches the ear of a listener, his eardrum oscillates at the same frequency.

The oscillations of a piano wire, the air, and the eardrum are all sound. The piano wire oscillates up and down; the air oscillates forward and backward; and the eardrum oscillates in and out, and we hear sound. Sound waves reside in the oscillation of the materials that generate it, propagate it and receive it. They have, however, an independent existence. Sound waves are like ordinary matter. They consist of energy, momentum and information. The oscillation of the air consists of air molecules moving backward and forward, and hence contains energy.

Sound waves move in a definite direction at a definite speed. In air the speed is about 300 meters per second. A strong sound wave, say one created by a supersonic jet, can break a window. So a sound wave can be like a moving hammer. It carries momentum. Sound waves also contain information. We use sound to convey meaning. So a sound wave contains energy, momentum and information.

Sound, on the other hand, is not the piano wire. It is not air. It is not the eardrum. It does not consist of atoms and molecules. We cannot hold sound in our hand. We cannot weigh sound on a scale. But sound is real. It can be heard. It can be measured with an instrument. Its existence is undeniable.

Q2.2.4. What is the smallest lump of energy of a sound wave?

When we hit a piano wire with a piano key, a sound wave is produced. A weaker and weaker sound is produced if we hit the piano key softer and softer, until it cannot be heard. The intensity of the sound is proportional to the strength we use to hit the piano key. It is natural to assume that the energy of the sound wave is continuous. However, quantum theory says that if the energy of the sound gets small enough, this idea does not apply.

Energy is not continuous. It comes in lumps, or packets. The smallest lump of energy of a sound wave is called a phonon. In quantum physics, a sound wave is a batch of phonons. Phonons are the smallest lumps of energy of oscillation that comprise a sound wave. They are quanta of energy. Phonons carry energy and momentum. The precise mathematical formulation is called quantization (discussed in Chapter 5).

When a baseball flies into the air, its energy and momentum is the sum of the energy and momentum of all the molecules that comprise the baseball. When we speak, sound waves come out of our mouth. The energy and momentum of the sound is a sum of the energy and momentum of all the phonons that make up the sound. Phonons are oscillations of the air. Take away the air, and we do not have air molecules. There are no oscillations of molecules. There are no phonons, and hence no sound. Sound cannot propagate in a vacuum. Phonons are not air. Stationary air does not make sound. Phonons have independent existence apart from the media they propagate in. Their existence is on a par with the existence of electrons and protons.

Qi is the oscillations of meridians. Qi carries energy, momentum and information. It also has the smallest lump of energy. We shall call such an object a qion. Qi then consists of qions. Qions exist independently of meridians.

Q2.2.5. When a piano wire is stretched in place inside a piano, we say it is in the ground state. Does it have any energy?

A stretched piano wire inside a piano is in an ordered state. It is the ground state, the lowest energy state of the wire. The energy is not zero, however. Naively, one assumes that the oscillation of a piano wire after being hit by the piano key will get smaller and smaller as it dissipates its energy through emitting sound waves to its surroundings, and that eventually the oscillation of the piano wire will stop. It turns out that this is not true. The lowest energy state of a stretched piano wire is finite, and is called zero point energy. Even at zero point energy, oscillations continue, inasmuch as all the particles of the atom are alive with energy, moving in relation to each other. (Fig. 2.2.5).

When meridians are at their lowest energy state, the ground state, they also have zero point energy, and are alive. Through deep meditation, we may arrive at such a state, where our mind and body are completely still. Although we are not thinking or moving physically, it does not mean there is no motion. We just are not using the energy to create. Our energy is in a state of coherence.

Q2.2.6. We can hear sound, but we cannot hear or see qi. How do you explain that?

Sound waves are present even when we cannot hear them. For example, ordinary table salt is made up of sodium chloride crystal. At room temperature the salt particle has thermal energy. This energy is the sum of kinetic energy of the sodium and chlorine nuclei inside the salt crystal. The back and forth oscillations of these nuclei occur with a definite frequency. These oscillations are phonons, which are the particles that make up the sound wave. The collective motion of nuclei in crystals are generally treated as a sum of phonons, which all have different energy and momentum.

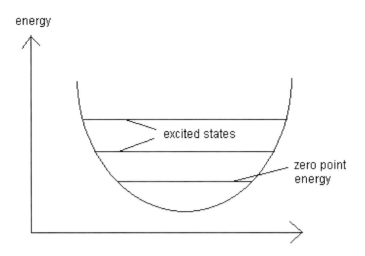

Fig. 2.2.5. Energy states of an oscillator. the lowest energy state, the ground state, is called zero point energy. There are many excited states.

Let us now use the simplest approximation for meridians. Meridians in their lowest energy state are made of water clusters with a permanent dipole moment arranged in a continuous chain-like manner. This is the ground state of a living meridian. To illustrate, when a qigong master decides to die, he breaks up his meridians, which results in death. When the water clusters are random and disordered, they can no longer carry qi. But, as the simple constituents of meridians, they are expected to exist even when there is no life in the meridians.

The highest level of existence for meridians is when there are collective excitations of the water clusters. There could be many collective excitations. The simplest kind would be the compression and expansion of water clusters along the meridians. These excitations are what we call qi. In complicated structures like water clusters with permanent electric dipoles, it is possible to have many kinds of collective excitations. They could be phonons, polarons, solitons, or some as yet undiscovered quasi-particle states.

A visual way to imagine the many different kinds of collective motions that a chain of clusters can have is to think of a key chain. You can rotate the chain, in helix fashion, similar to turning a screw. The chain could turn clockwise or counter-clockwise. You could shake it up and down. You could twist it, etc.

Qi is identified with these collective motions. There are many different kinds of qi. The smallest lump of energy of qi we will call a *qion*, a word obtained by adding on to qi. We have coined this term in the same tradition as are the terms photons and phonons. Photons are the smallest quanta of electromagnetic radiation. Phonons are the smallest quanta of sound. Qions are the smallest quanta of oscillations along meridians, but there are many different kinds of oscillations, which we shall discuss later. Qions are quanta of life.

Q 2.2.7. Qi is important in acupuncture. How are qi and acupuncture related?

Qi, the oscillation of meridians, carries energy and information. The information, most likely, is carried through its frequency and amplitude. In Chapter 1, section 1.2, we have listed the experiments that show different frequencies of electro-acupuncture have different effects on the body. In acupuncture there is a standard way of enhancing a deficiency or reducing an excess by using different ways of twisting the needle. It is likely that enhancing a deficiency involves causing an increased amplitude in oscillation of the meridians, and reducing an excess involves causing a decreased amplitude of the oscillation.

There are different kinds of oscillations of meridians. Twisting the needle in different ways during acupuncture most likely stimulates and resonates different kinds of oscillations in the meridians, and so can produce different effects on the body.

Understanding different kinds of oscillations of meridians with various frequencies and amplitudes would, no doubt, lead to more efficient and effective acupuncture methods.

Life – Creation of qions. Death – destruction of qions

Q2.3.1. What is life?

> When a baby is conceived, life is created.
> When a seed sprouts, the life of a plant begins.
> Human life is created from the elements of the earth.
> The absence of life is death.

Creation is the essence of life. Without creation there is no life. Only one theory in physics deals explicitly with creation – the quantum field theory. A quantum field contains operators, or elements, that can create objects from out of a vacuum. It also contains elements that can annihilate objects. The unity of opposites – creation and annihilation – which is used to explain life and death, is the critical element in defining the concept of the quantum field.

The annihilation of life is death. The opposite of death is birth. Death is as important as birth. Everyone faces death in the end. Life consists of birth and death, or creation and annihilation.

Q2.3.2. What is the basic essence of life for a complex, multi-leveled human being? Is there one system that contains the life of the entire human body?

Life exists on every level of a human being. The heart has life. The brain has life. There are different views as to what defines the absence of life, or death. Some people regard the cessation of the heartbeat as the moment of death. When a person dies, the various electrical oscillations of the heart, which can be recorded by an electrocardiograph (ECG), stop.

Some people regard the cessation of brain waves as the moment of death. When someone dies, there are no more electrical brain waves.

Oscillations of the organs, such as the heart and the brain, are vital signs for these organs. Cessation of such oscillations indicates death.

In traditional Chinese medicine, meridians are regarded as the overall system. So oscillations, called qi, in the meridians, are vital signs. If there is no oscillation, or qi, in the meridians, the person is considered dead.

Quantum fields contain waves that describe oscillations. So quantum fields are the natural overall system for describing the presence of life in a human being.

Q2.3.3. Quantum fields are quite abstract entities for most people. What advantage do we gain by using quantum fields to describe qi?

The meaning of qi has puzzled many people who want it explained scientifically. By using quantum fields to describe qi, we provide a scientific explanation of qi. This can be tested in future scientific experiments. At the moment this is still a hypothesis, but it is a hypothesis that can be proved or disproved.

Q2.3.4. We've said that life may be described by quantum fields. Can you give a couple of examples of the use of quantum fields to describe physical phenomena?

One example is sound. If we hit a piano key, the piano wire oscillates. We have created a sound with a specific frequency. The piano wire then makes the air oscillate back and forth. When the oscillations of the air reach the ear, the eardrum starts to oscillate. The nerves in the ear transmit this oscillation signal to our brain, and we then hear sound. The sound is created in the air when we hit the key, and it is destroyed when the air loses its energy to the eardrum.

Sound is made of waves. Sound waves may be oscillations in a wire, in the air, or in the eardrum. The smallest particle of energy of a sound wave is called a phonon. Phonons may be represented by a quantum field.

Another example is light. If we have a light bulb hanging from the ceiling of a room and we want to have light, we turn on the switch. The switch turns on an electric current, and the electrons in the electric current in the wire of the light bulb create light.

Light is made of electromagnetic waves, which are oscillations of electrical and magnetic fields. They propagate through the air and reach our eyes. The retinas in our eyes send a signal through the nerves to our brains. Then we see light. Light is created at the wire of the light bulb, and is annihilated when it reaches the retina.

The smallest particle of energy of light is called a photon. Photons may be described as a quantum field. Quantum fields can represent both phonons and photons. There is a quantum field for each.

A quantum field of phonons can create or annihilate sound. When we describe sound wave propagation in air, we use a quantum field of phonons. A quantum field of photons can create and annihilate light. When we describe the creation of light in a light bulb, we use a quantum field of photons.

There are oscillations in meridians, which we refer to as qi. The quantum fields of qi can create and annihilate oscillations in meridians. When we describe the behavior of oscillations in meridians, we shall use quantum fields of qi.

Q2.3.5. How much are quantum fields being used in physics? According to what principle do they interact with one another?

Let us first give a practical example. When we watch television, the images on the screen are carried to our eyes by photons. The photons are created when the electrons from the electron gun inside the television strike the atoms on the screen. The production of photons through collisions of electrons with atoms is calculable using a principle called the action principle. Within the action of the action principle is the precise form of how a quantum field of electrons interacts with a quantum field of photons. The action principle enables us to derive precise laws that govern the motion of electrons and photons. The application of these laws to a given initial state of electrons and a given final state of photons will yield the probability and other information we need to design the television screen.

Quantum fields are used extensively in many branches of physics, including nuclear physics, particle physics, solid-state physics, laser physics, astrophysics, etc. Quantum fields interact with one another according to principles and established rules. The action principle is one of the most important of these principles. It so named because actions follow from it. The way particles or waves move is determined by the way quantum fields interact with one another, as given in the action of the action principle.

The action principle and explicit forms of quantum fields with their precise mathematical forms are presented in Chapter 5.

Q2.3.6. How are quantum fields being used in qi, organs and meridians?

In a living person, qi circulates along the twelve meridians 24 hours a day. Suppose a person has a stomach ulcer. The stomach sends a signal to the stomach meridian and influences the oscillations (qi) in the stomach meridian. Qi in the stomach meridian will carry the information that the stomach is not well, and circulate the message to other parts of the body.

This may explain why, by feeling the pulse, a good Oriental doctor can determine the well being of a person's stomach. The doctor then may insert needles in acupoints in the spleen meridian. By the stimulation of qi in some acupoints on the spleen meridian, signals will be sent to other meridians, such as the stomach meridian. So the qi in the spleen meridian interacts with the qi in the stomach meridian. After this interaction, signals are sent to the stomach and other parts of the body to start curing the stomach.

The interaction of qi with qi is described by the interaction of quantum fields. While we know little about the interaction of qi with qi inside the human body, we understand well the interaction of quantum fields in other parts of the universe outside of human beings. The knowledge of quantum fields in other areas could be brought to understand the interaction of qi inside human beings.

In general, qi balances and coordinates the functions of various systems and organs. Quantum fields represent qi that circulates in meridians. Meridians pass through various organs in the body. The various organs send signals encoded in different forms of oscillations (qi) to the meridians, and meridians carry signals (qi) to various organs. Meridians are the place where various oscillations interact. The quantum fields of oscillations (qi) in the meridians interact with one another according to some action, which we do not need to know precisely.

The action principle may be regarded as the purpose of the person, or the promotion of his total well-being. Quantum fields (qi) interact with one another and then are sent to various organs to regulate them. The regulatory function of qi is to balance the functions and activities of various organs so that they work in harmony for the well being of the entire human being.

The function of acupuncture is to change the oscillations in the meridians so as to balance various organs. If one organ is hyperactive and overworked, acupuncture on acupoints on the meridians will slow it down. If another organ is depressed, and slows down, acupuncture will activate it more. Meridians are the places where all input signals (qi) meet, interact, and come up with a solution. The signals are then circulated back to the organs so that the organs work harmoniously. Meridians and qi coordinate the function of various organs.

Q2.3.7. What do you hope to achieve with a quantum field theory of meridians and acupuncture?

Brain waves are necessary for life.
Heartbeats are necessary for life.
Qi circulation in meridians is necessary for life.

A quantum field theory that describes waves, beats, and qi in brain, heart, and meridians will be able to explain important aspects of life.

We have outlined here a very rudimentary framework for a quantum field theory of qi, meridians and acupuncture. It can be used to perform qualitative analysis. We even have some predictions that we shall discuss in Chapter 4.

What is the chance for success? Our experience with the quantum field theory of electrons and photons is very encouraging. The quantum field theory of electrons and photons explains all electromagnetic phenomena. There is no exception. The accuracy of some experiments in quantum field theory is phenomenal. The theory predicts that the mass of a particle is equal to the mass of its anti-particle. Experimental findings have verified this prediction of equality to one part in a hundred trillion. It is like calculating the gross national product of the United States and verifying its accuracy to one cent in one trillion dollars.

We hope that quantum field theory will eventually explain major aspects of life phenomena as we now know them.

On the practical side, using the qualitative guidance of quantum theory, we may be able to construct a much more effective and versatile acupuncture method for curing a greater variety of diseases and health problems. We may be able to construct a much more accurate diagnostic method as well. (See Chapter. 4.) Patients and the general public may reap the practical benefits of this theoretical development.

Meridians – the primitive site of life

Q2.4.1. Do plants and animals have something bigger than cells in common? Is there anything above the cellular level that is shared by all living organisms?

When I studied biology in high school, I learned that the biggest small unit that all living organisms had in common was the cell. Living organisms did not share anything above the cellular level. We have a brain, liver, kidney, etc., but plants do not have such sophisticated organs. Plants have leaves, trunks and roots, which we do not have. Higher animals have nervous systems, hormonal systems and blood circulatory systems to connect various parts. A meridian system coordinates all these other systems. Plants do not have a nervous system, hormone system or blood system. The size of the roots and branches determines the size of the plant, so that it doesn't grow too big or remain too small. Various parts of a plant work together to make each plant unique. It is possible that plants also have a coordinating system like a meridian system. The meridian system, then, may be one that is shared by animals and plants alike.

Q2.4.2. Is there any evidence that plants have meridians?

There are many investigations into the possibility of the existence of plant meridians (some of which are listed in Table 2.3.1). The subjects of study are the soybean, pole bean, and philodendron. The research workers have found that the plants have low resistance points, and needles at these points reduce the electrical resistance of the main vein. Low electrical resistance is the hallmark of meridians.

Inserting needles in a soybean plant was seen to increase the temperature by half a degree Centigrade. Furthermore it was observed that leaves of philodendron will emit very low frequency sound waves when a laser stimulates them. The frequency range is from 50 Hz to 120 Hz. This is the same range of frequency that is effective in electro-acupuncture for humans.

Highlights of Table 2.3.1. Meridians in Plants – meridians as most primitive and fundamental system

- Two needles were inserted into two low resistance points of leaves of soybean plants and the electrical resistance on the main vein was reduced by 26%.
- Laser beams shining on leaves of philodendron induced them to produce sound in the range of 50Hz to 120Hz.
- Two needles inserted on opposite sides of the stem of unifoliolate buds of pole bean and bush bean increased photosynthesis rate by 20.5%.

Q2.4.3. How is the meridian system related to other systems of the body?

Dr. Shang's theory, discussed in Chapter 1, says that the meridian system is the most primitive system. It is first developed to coordinate various parts of a living organism, when a single cell evolves into a multi-celled organism. Later more sophisticated and elaborate systems of nerves and blood vessels evolve. Of all systems, the meridian system is the most primitive, and hence also the most fundamental. The meridian system coordinates all other systems to make the body function as a whole.

Q2.4.4. Is there any evidence that the meridian system is the most fundamental system?

As we discuss in Chapter 1, a human body has a multi-leveled structure. The lowest level consists of positive and negative charges and photons. The highest level consists of systems. Acupuncture works on the lowest level – the positive and negative charges of the polarized water clusters. It influences the oscillations of the meridians. The meridian system influences all levels of the human body: systems, organs, cells and molecules.

Much research has been done on acupuncture at the Zusanli, ST36 acupoint of the stomach meridian in rats. Acupuncture affects all levels in rats – systems, organs, cells and molecules.

- Using the radio-immuno-assay technique, it is found that acupuncture at ST36 influences the gastro-intestinal system.
- It influences the immune system by measuring Natural Killer (NK) cytotoxicity.
- It influences the blood circulatory system of the pancreas.
- It influences the hormonal system, such as the adrenal glands.
- It influences the functions of organs – the duodenum, jejunum, ileum,

spleen, liver, brain and pancreas.
- It influences cells, as in the discharge of neurons in the lateral hypothalamic area of the brain.
- It influences molecular layers of the body, resulting in an increase of insulin, beta-endorphin, interleukin –2, and interferon gamma.

(The experimental data is given in Table 2.3.2.)

The explanation is quite simple. Stimulations at acupoint ST36 will influence the oscillations of the stomach meridian, which will then propagate to different parts of the body that change the behavior of different physical levels. Remarkably, a single needle at a prescribed point on the skin can affect all physical levels in an animal.

Highlights of Table 2.3.2. Various effects of stimulating ST36

- It increased high levels of interleukin (IL)-2, and interferon (IFN) gamma, and enhanced splenic natural killer (NK) cell cytotoxicity.
- Electro-acupuncture (EA) at ST36 depressed high blood pressure and blood hyperviscosity, which may be mediated by activation of GABA receptor in the brain.
- By influencing unit discharges of neurons in feeding center of lateral hypothalamic area, it may abolish the inhibitory reaction induced by distending the stomach.
- By measuring quantitative histochemical changes of glycogen SDH, EA is shown to improve the function of the adrenal cortex.
- Acupuncture is effective in treating liver injury induced by carbon tetrachloride in rats.

This evidence suggests also that there is a much different type of water cluster at acupoints. Upon stimulation, these water clusters generate electromagnetic waves, electric fields, and/or sound waves. They oscillate at different frequencies. These waves travel through meridians to various organs that resonate with similar water

clusters in that organ. Some waves may affect nerves, which will alert the relevant part of the brain for appropriate action.

Q2.4.5. Is there any data on the effect of acupuncture at ST36 on a human body?

We expect the effect of acupuncture on humans will be similar to that on animals. Obviously it is not easy to experiment on a human brain or spinal cord. However, we also have a different kind of data. In Picture 2.3.1 we show a thermograph of a patient. The patient reported feeling dizzy. We took a picture of his face. It showed hot regions around the eye and mouth, with maximum temperatures at 36.2^0 C, and

36.02° C (in white) respectively. Since the hot area (in red) around the mouth follows the stomach meridian, I asked him if his stomach was unwell. He said that it was. I decided to use one needle on the acupoint ST36 Zusanli, which is just below the left knee. In 14 minutes, another thermograph was taken. The pattern of color on his face had changed. The white spots around the eyes and month had disappeared. Quantitatively, the maximum temperatures dropped to 35.76° C and 35.4° C respectively. This is a decrease of 0.44° C and 0.62° C. The decrease is significant, because the infrared camera is accurate to 0.01° C. When I asked the patient what he felt, he replied that he felt better.

Inserting a needle at a distance from the area of complaint would be effective, according to an acupuncture textbook, as long as the same meridian passes through both areas. This assertion is proven by these two thermographs.*

Q2.4.6. Can you describe briefly how acupuncture cured sickness in this case?

When a needle is inserted into the muscle at acupoint ST 36 beneath the left knee, the muscle will contract at its natural frequency. This influences the qi, or the oscillations, on the stomach meridian. It is like a ripple on a canal. Such a ripple will travel down along the stomach meridian, up the leg, through the stomach area, to the face. The ripple affects the inflamed area on the face where the stomach meridian passes through. This is similar to how a radio receives a signal from a broadcasting station. A resonance effect occurs to start and speed up the self-healing process. Temperature is reduced in the inflamed area of the face.

Q2.4.7. What are oscillations on the meridians?

In nature oscillations occur continually. The most sophisticated concept invented thus far to describe oscillations in a fundamental way is the quantum field. While light is an oscillation of an electromagnetic field, its smallest lump sum of energy, called a photon, is described by a quantum field. Sound is the oscillation of particles in air, liquid or solid. Its smallest lump sum of energy is called a phonon, which is also described by a quantum field. The oscillations of meridians are most likely complex, and consist of distinct different kinds of oscillations. The smallest lump sum of energy associated with all the oscillations of the meridians we call qions.

The simplest kind of oscillation is described by a plane wave, which has a definite frequency, wavelength and amplitude. The sound produced by a tuning fork that a technician uses to tune a piano is closest to a plane wave. Hitting a piano key will produce a sound, which is a sum of plane waves, with a small spread of frequency and wavelength.

There are more complicated oscillations, which cannot be expressed as a sum of plane waves. One of these is called a soliton, which is a solitary wave that propagates forward without diminishing its intensity. An example of a solitary wave is a big ripple in water traveling forward in a canal. Qi can have a solitary wave component, which travels along the meridians. Solitary waves have qualitative properties that can be felt as qi moving along meridians after a needle is inserted in an acupoint.

For oscillations in the meridians, there are probably plane waves as well. Simple plane waves have a definite frequency and amplitude. In electro-acupuncture, the electric stimuli have definite frequencies, as we discuss in chapter one. Different frequencies have different effects on the body. For normal functioning of the body, there must be a suitable intensity of qi circulating along the meridians. The intensity of a plane wave is the square of its amplitude. If the intensity is more than that of the normal oscillation, there is excess, which has to be reduced in acupuncture treatment.

If the intensity is less than that of the normal oscillation, there is a deficiency, which needs to be enhanced by acupuncture.

Q2.4.8. What effects do these waves in meridians have?

There are two kinds of effects the waves have on their surroundings. One is the resonant effect; the other is the non-resonant effect.

The resonant effect of sound is sometimes employed to amuse guests at a party. Someone skillful can wet the mouth of a glass and slowly rub it to create a natural vibration of the glass. An identical glass, placed farther away, will resonate and break. Similarly, when soldiers march on a bridge, it is important that they do not march in time with each other. Otherwise the matching frequencies of the soldiers' steps might happen to match the natural frequency of the bridge. The bridge would then start to swing and break. The sound of a bomb blast can also break glass, but this is not a resonant effect. This works through the sheer pressure inherent in the sound wave.

Electromagnetic fields also have resonant and non-resonant effects. The electromagnetic fields emitted from a broadcasting station are received by radios many thousand miles away through the mechanism of resonance. The frequency of vibration is important. Radios must be tuned to the right frequency in order to receive the correct signals. The electromagnetic field of an electric motor, on the other hand, does not work on the principle of resonance, but on the principle of generating mechanical force by passing electricity through a magnet.

It is most likely that oscillations of the meridians also have resonant and non-resonant effects. The resonant effect will travel far from the acupoints, whereas the non-resonant effect probably has an effect near the acupoints. The non-resonant affect might take the form of affecting some body fluid, as in an experiment using acupuncture at the urinary bladder meridian to induce urine. Since oscillations of polarized clusters will naturally generate electric fields, it is quite possible such electric fields will stimulate the nerve endings near the acupoint, and signals will then be sent to the central nervous system. This kind of effect is also a non-resonant effect.

Q2.4.9. In traditional theory there is a qi that circulates around the twelve meridians on a daily basis. What would it consist of?

In our model it is most likely that the daily qi that circulates around the twelve meridians, as shown in Fig 2.4.9, consists of ionic qions. The photonic qion travels at the speed of light. The phononic qion travels at the speed of sound, which is about 300 meters per second. So it does not take more than a few seconds for both photonic and phononic qions to travel around all twelve meridians. Ions travel at a much

slower speed, which could be measured by a radioisotope experiment. Then one could easily verify whether these ions complete the whole cycle of twelve meridians in one day.

Clicking the mouse

Q2.5.1. The brain is regarded by most scientists as the site of high-level activities. How are meridians related to the brain?

According to Dr. Shang's evolutionary theory of meridians, meridians evolve as the earliest system. They have been part of living organisms for many hundreds of millions of years before the evolution of the brain. They are supposed to be evolves, branches from the meridians evolve to connect to it. There would be no new meridians added. Since the brain is fast and efficient, it is natural that meridians would use the brain to coordinate various body functions. This has been demonstrated by many scientific experiments and clinical tests (Table 2.4.1).

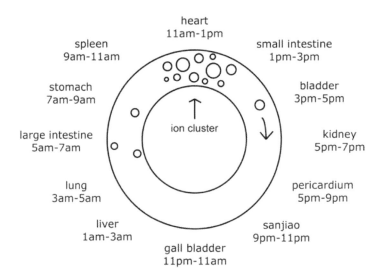

Fig. 2.4.9. A cluster of ions takes 24 hours to rotate around all meridians.

Acupuncture at various points of different meridians affects different parts of the brain. For example,

- Acupuncture at acupoints on the legs, such as GB37 Guangming of the gallbladder meridian and Zhiyin (BL67) of the urinary bladder meridian, affects the eyes. Scientific experiments suggest that these effects go through the visual cortex of the brain.
- Acupuncture of the stomach meridian at acupoint Zusanli (ST36) on the leg can alleviate immobilization stress in rats, but the effect goes through the hippocampus in the brain.
- Acupuncture at acupoint Shenmen (HT7) of the heart meridian may treat diseases associated with maternal separation of the young. The effect is seen on the dentate gyrus.
- Acupuncture at Shaofu (HT8), Zutonggu (BL66), or Xingjian (LR2) has antipyretic effects on fever in rats. The cytokine expressions in the hypothalamus are affected, so the liver meridian also affects the brain.
- Electro-acupuncture at Neiguan (PC6) and Jianshi (PC5) has an inhibitory effect on the reflex autonomic response in rats. It is related to activation of mu- and delta-opioid receptors in the rostral ventrolateral medulla (rVLM).
- Acupuncture at Hegu (LI4) has prominent analgesic effects. It is related to activation of the hypothalamus with an extension to the midbrain, the insula, the anterior cingulated cortex and the cerebellum.
- Using functional magnetic resonance imaging (fMRI), it is found that needle manipulation on Hegu (LI4) produces prominent decreases of fMRI signals in the nucleus accumbens, amygdala, hippocampus, parahippo-campus, hypothalamus, ventral tegmental area, anterior cingulated gyrus, caudate, putamen, temporal pole and insula in normal healthy subjects. Signal increases were observed primarily in the somatosensory cortex.
- From the study of manganese-enhanced functional magnetic resonance imaging on New Zealand white rabbits, it was found that stimulation on Zusanli (ST36) resulted in activation of the hippocampus, whereas stimulation on Yanglingquan (GB34) resulted in activation of the hypothalamus, insula and motor cortex.
- Electroacupuncture at Baihui (DU20) and Rengzhong (DU26) of normal rats activates the survival Akt signals in the cortex, caudate, CA1 sector and dentate gyrus of the hippocampus and ventral posteromedial thalamic nucleus.

Remark on thermographs: A thermograph of a patient is taken by capturing the infrared radiation emitted naturally from warm human skin with an infrared camera. Then a computer uses this information to calculate temperatures of each point on the surface of the body. The temperature distribution is displayed in 16 shades of color, in increments of 0.5^0 C, for a range of 8^0 C between the maximum and minimum temperature displayed. White is the hottest, then red, orange, yellow, green, blue, and black, which is the coldest.

So scientific experiments have demonstrated that at least four yang meridians (stomach, urinary bladder, gallbladder and large intestine), three yin meridians (heart, pericardium and liver) and the Du meridian are related to the brain.

Q2.5.2. Computers are faster and more efficient than human beings in many applications. But computers obey humans. Is there any similarity in the relation between the brain and the meridians, and between computers and humans?

When I first learned computer programming at the University of Illinois, Urbana, as a student in the early 1960s, computers were just coming into use for ordinary people. I had to use paper tape and then paper cards for input and output. But the awesome power of the computer had already fired people's imaginations. Would human beings be superceded and rendered useless, because computers would eventually calculate faster and memorize better than humans? It hasn't quite turned out that way. Now that the time of enormous computer power has arrived, computers are serving humans rather than enslaving them. How is this possible? It is because humans retain final decision-making power over what computers do. With a personal computer, clicking the mouse decides what the computer does, and a human being is the one who clicks the mouse.

A personal computer is a rather sophisticated piece of equipment. Few people can claim to understand the intricacy of all the hardware and software that goes into it. But now computer-illiterate housewives are able to correspond with their children via e-mail with the use of a mouse. When we turn on the computer, it remains idle and does not know what to do. We have to click the mouse to give directions as to how the computation should proceed.

The computer is complex and does all the work quickly and thoroughly. The mouse in comparison is simple, unsophisticated, and does not do much work. But the computer listens to the command from the mouse. Clicking the mouse is done by a human operator, and so is linked to a higher level than the electrical activities going on among the chips inside the computer.

Similarly, a human brain is a rather sophisticated organ. No one has yet claimed to understand the detailed workings of the brain. It works quickly, via extremely complicated electrical signals.

In our view, meridians play the decisive role, similar to how a mouse communicates with a computer. Though the meridian system is primitive and works slowly, it makes the final decision as to how a human being should proceed. Making crucial decisions in life belongs to a higher function than that of electrical signals in the brain. It belongs to the spiritual level of human activities. In our view, spiritual activities come from the interaction of quantum oscillations in the meridians.

We use the word spiritual here in a very narrow sense, in the way it is used in the tradition of qigong masters. The spiritual state as defined here is closely related to supreme health. It's a state that opens one to higher intelligence and inspiration. It is characterized by ease in action, clarity of focus and intention, perseverance, non-attachment, resilience, openness, creativity, responsiveness and balance.

Since spiritual activities, as defined here, result from interactions among quantum oscillations, their properties will eventually be measurable by physical instruments. Activities that cannot, in principle, be measured by physical instruments or detectors will be excluded from our definition of spiritual activities.

Q2.5.3. How do we make decisions in clicking the mouse, and in directing the brain?

When we click the mouse to decide which program to run, the decision comes from our thinking process, which is quite independent of the computer's calculations and processing. Clicking the mouse depends on our consideration of various factors quite outside the scope of the computer. Similarly the interaction of quantum oscillations in the meridians is independent of the electrical signals of the brain. When the meridians in the system decide on the direction of human activities and coordinate various organs to proceed harmoniously, the decision comes from interactions of various kinds of quantum oscillations.

Q2.5.4. What is the mechanism for making decisions? How is it related to the balancing of yin and yang by the meridians?

Clicking a mouse merely requires moving our finger down and is simply a mechanical process. The circuit inside the mouse contains an on and off switch and other electric circuits. How does a particular nerve cell decide to fire? How do the interactions of quantum oscillations in meridians affect the firing of neurons? We do not know at present.

Since we use quantum fields to describe oscillations in the meridians, we shall use the latest results in quantum field theory concerning spontaneous symmetry-breaking as a model for decision-making. Take the example of a ball resting on the top of a hill. The height of hill is a measure of its potential energy. The ball is in an unstable equilibrium state. It has a perfect left and right symmetry. It can fall to the left as well as fall to the right. A tiny push on the ball, say, from the left, will break the left-right symmetry, making the ball fall to the right. Alternatively a tiny push on the ball from the right will make the ball fall to the left (Fig. 2.5.4).

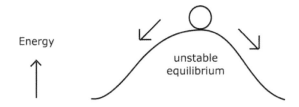

Fig. 2.5.4. When a ball rests on top of a hill, it is in an unstable equilibrium. A small perterbation will make it roll either to the right or to the left.

We assume this is a useful model for how we decide, for example, to turn left or to turn right when we drive a car to an intersection. Similar symmetry-breaking mechanisms will operate in more complicated situations involving the interactions of quantum oscillations and their effects on the functions of the body. Then left and right may take on a more general meaning. It might mean to do more or less. It may mean to have more yin, less yang, or more yin, less yang.

According to traditional Chinese medicine theory, one of the most elemental parts of being healthy is maintaining the balance of various functions of the body. When we eat, our stomach should secrete enough acid to digest the food. Too little acid won't digest the food, and the food will be wasted. Too much acid will harm the stomach itself and can eventually cause an ulcer. Similarly if there are attacks on the body by outside foreign bacteria, there should be enough antibodies to fight them. It there are too few, the bacteria will multiply, and we will get sick. If there are too many, immune system is in trouble. In abstract terms, to remain healthy, the yin and yang of the body must be balanced.

Balancing the body function is required when there is too much of one thing in the body, such as acid or antibodies. A decision to reduce the excess is made through the interaction of quantum oscillations in the meridians. If there is too little of something, then a decision is made to increase it.

When a human body loses the balance of yin and yang, the person becomes ill. Acupuncture at the right acupoints on the right meridians will restore the balance of the body, and the person will be cured and become healthy again. See, for example, in Q2.3.5, our discussion of how acupuncture at ST36 on the stomach meridian cures digestive problems. We have also listed the latest research results of its various effects in Table 2.3.2 (see Q2.4.4).

Meridians are not separate from each other. They form a network, split out into smaller and finer branches. The branches are further split into capillaries, similar to how the blood system branches. So meridians do not just balance the main organs. The effect of acupuncture on acupoints on the meridians carries the oscillations along the meridians to various branches and capillaries, and can hence influence various parts of the body, including tissues and cells.

In our theory meridians are made up mostly of stable water clusters with permanent dipoles. So these are what make up the branches and capillaries of the meridian system. An infrared picture can show within minutes the changes due to acupuncture in a large area of the body, not just confined to a narrow strip along meridians.

This evidence suggests also there is a much different type of water cluster at acupoints. Upon stimulation, these water clusters generate electromagnetic waves, electric fields, and/or sound waves. They oscillate at different frequencies. These waves travel through meridians to various organs that resonate with similar water clusters in that organ. Some waves may affect nerves, which will alert the relevant part of the brain for appropriate action.

Q2.5.5. Is there evidence for the balancing effect of acupuncture?

The effect of balancing the left and right sides of the body using acupuncture on one acupoint on one meridian was demonstrated in a little experiment we performed. My wife had some pain in her left leg and felt uncomfortable. The pain did not go away by itself for several days. An acupuncturist friend of mine visited us, and we asked him to treat my wife's pain. Before the treatment we took infrared pictures of both of her legs, shown in Picture 3.3.1. From these two pictures, it was quite clear that the color distributions of the two legs were different. Since different colors represent different temperatures, this meant the temperature distribution of the legs was not the same, or the two legs were not balanced.

The acupuncturist looked at the pictures and said that the hottest spot was closest to the liver meridian. He decided to insert a needle at LR2 Xingjian of the liver meridian on the left leg only. I suggested only one needle should be inserted on the left foot away from the hottest spot, so that the effect would be clear. The needle was barely observable as a blue image in Picture 3.3.1, top picture. Infrared pictures were taken again after ten minutes of acupuncture. In the bottom picture, we see that the color distributions of the two legs approach each other – the temperature distribution of the two legs became similar, and they had become balanced again. The pain went away after the treatment and did not come back to bother my wife again. Similar balancing effects were observed in other cases we treated.

Q2.5.6. How do meridians interact among themselves?

There are fourteen meridians in the human body: twelve regular meridians and the Du and Ren meridians. These meridians carry qions, circulating around the body. They are like fiber optic tubes carrying photons. Meridians, we believe, are also where qions interact with one another. Qions are the smallest quanta of oscillations occurring in the meridians. When they interact, they modify and influence one another. They interact in meridians, and have boundaries defined by the boundaries of the meridians.

When the natural process of interactions and circulation of qi is blocked or disturbed, the person will be out of balance and not feel well. Stimulation from acupuncture will accelerate or decelerate oscillations in the meridians, so that there are a larger or smaller number of qions circulating in the meridians. Interactions among qions will become stronger or weaker, and the internal organs of the person will become balanced again.

In traditional meridian theory, there are elaborate rules governing the type of relationships in the interactions of meridians and organs: mother-son relationships, surface, and internal relationships. These rules come, no doubt, from the accumulation and summary of a vast body of clinical experiences by a large number of practitioners over a long period of time. Here our quantized version of meridian theory attempts to provide a framework so that these empirically observed relationships may potentially have a solid theoretical foundation in physics, and that eventually the consequences of these relationships will be measurable by physical instruments that detect qions circulating in the meridians.

Let us elaborate more on the concept of balance. Normally in physical science or engineering practice, one always pursues an idea to the ultimate breakdown point. When we build a radio, we like to build one with the largest possible power output and with the best detecting efficiency. A human does not work that way. We do not push our heart to pump as fast as it can, with the highest pressure, so that it circulates the largest possible amount blood to every part of the body. In fact, when a foreign body invades our body, if we overreact and produce too many antibodies, as in the case of hay fever, we suffer.

For a living organism to work optimally there must be a balance: blood pressure must be neither too high nor too low; heartbeats must be neither too fast or too slow, and there must be neither too many nor too few antibodies. Skin temperature should generally be neither too hot nor too cold, because too hot or too cold in any given spot generally means dysfunction of some part of the body.

Each organ of the body has a normal optimal state of functioning. For each task we perform, there is also an optimal point of correlation of different organs. For example, in running, we require more oxygen and more energy supply to the muscle. The heart cooperates by beating more and the lungs breathe more. It is believed that qions from the heart and lungs interact to create a balance.

In current Western medical theory one assumes such coordinating functions are done automatically through the nervous system. Traditional Chinese medicine theory does not exclude the regulatory function of the nervous system. The qion interactions make decisions for the nervous system, which coordinates and regulates the various functions of different parts of the body. This is analogous to how a human tells a computer what to do, but the microprocessor coordinates the function of various parts of the computer and its accessories, such as printers, monitors, modems, Internet access, speakers, etc.

Q2.5.7. How many types of qions are there?

Let us use the current results from quantum field theory and apply them to meridians and qi. There are two types of particles in nature, as we discuss in Chapter

1. There are electrons and protons that are fermions and have an anti-symmetric wave function. They obey the exclusion principle. No two electrons can occupy the same state. There are photons and phonons, which are bosons, and have symmetric wave function. They do not obey the exclusion principle, and many particles can occupy the same state. They have a cooperative effect. Similarly there are most likely two types of qions. One type of qion is a fermion, which we call an f-qion. The other type of qion is a boson, which we call a b-qion.

Some oscillations in meridians do not involve transport of matter from one place to another, but consist of periodical movement of clusters locally. Some of these oscillations may be periodical compression and expansion of local molecular clusters. Then they are phonons, which are quanta of sound waves. Some of these oscillations are infrared light transmitting along the meridians, as we discussed in Chapter 1. Hence b-qions consist at least of phonons and photons. B-qions may also consist of solitons, polarons, or other yet to be discovered objects.

Oscillations in meridians that actually involve transport of matter from one place to another are most likely f-qions. The external qi emitted by a qi-gong master has been measured as consisting of matter (the data for which will be presented in Chapter 4). So this external qi is composed of f-qions.

Q2.5.8. How do you explain the mother-son relationship among meridians? How do you explain the five elements – metal, wood, water, fire, and earth – in traditional meridian interactions in quantum theory?

Several thousand years ago, when meridian theory was formulated, it appealed to the laws of nature. At that time it was thought that there were five fundamental elements in nature: fire, water, wood, earth and metal. Since everything in nature is composed of these, it was thought that these five elements must play similar roles inside the human body. Thousands of years of experience seem to validate the idea that meridians interact among themselves in a fashion similar to how the five elements interact among themselves.

Traditional Chinese meridian theory has a set of mother-son rules, modeled after rules in nature, to describe the overarching system of interactions among meridians. Briefly, the rules governing the interaction of the five organs, or their five meridians, are modeled on the five directions in nature – east, south, west, north and the central; and on the five natural elements – wood, fire, metal, water and earth. As there are five organs – liver, heart, lung, kidney and spleen, the five directions and five elements can be matched to the five organs (Fig. 2.5.8). For example, water is the life of a growing tree, and so the kidney-water meridian serves as mother meridian to her son, the liver-wood meridian. If liver son is weak, then we must strengthen kidney mother in acupuncture. The relations among the five elements serve as a memory device for remembering the relations among meridians and are very useful to acupuncturists in their diagnoses and treatment of diseases.

We still base our meridian theory on laws of nature. As we describe in Chapter 1, the modern theory of matter is that it is made up of atoms and molecules, with their interaction described by quantum theory. Atoms and molecules interact through the exchange of photons, which are described by quantum fields. So in our modern quantum theory of meridians, meridians interact among themselves through the exchange of qions, which are described by quantum fields. The relations among meridians, intrinsically correlated with the five elements of fire, water, earth, wood and metal, will be derived in the future from the properties of interaction among the qions.

Sophisticated instruments have been invented in the field of particle physics and solid-state physics to detect and measure various kinds of particles described by quantum fields. When these devices are adapted to detect quantum field objects like qions in body, we expect many new results will be discovered, together with experimental confirmation of results of traditional meridian theory.

Let us clarify further the similarities between the fundamental processes of interaction of meridians and the fundamental processes in nature. When two opposite charges attract each other by electric force, the fundamental process is an exchange of photons between the two charges. When two massive objects like the

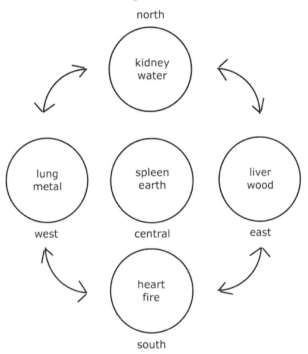

Fig. 2.5.8. Traditional Chinese theory associates five organs or their m with the five elements: water, fire, wood, earth and metal; and five pc

sun and earth attract each other by gravitational force, the fundamental process is an exchange of gravitons. Inside the atomic nucleus, the protons and neutrons stay together by exchanging mesons. So any fundamental process is due to the exchange of some particle.

It is natural, then, to propose that the fundamental process for the interaction of meridians is due to the exchange of qions. For example, salt is held together as a solid by the electrostatic force between its sodium and chloride ions. The attracting force involves the process of photon exchange between the sodium and chlorine ions. Similarly, meridians cooperate to harmonize the function of different parts of the body by the exchange of qions.

The elementary process of the electrostatic forces between sodium and chlorine ions come from the interaction between the extra electron in the negatively charged chlorine ion and the positively charged protons in the sodium ion, with the exchange of photons between the electron and the proton.

One of the elementary processes of the interactions among meridians similarly might be between two f-qions, one from each meridian, with an exchange of a b-qion. Interactions of the five elements, mother-son relationships, et. al. can be explained by the interactions of the qions in those meridians. The interactions of the qions are explicit in the quantum field theory, as discussed in Chapter 5, Q5.6. Those interactions have measurable consequences, which could be measured using modern scientific equipment. Details of simple mathematical interactions among qions are discussed in Chapter 5, Q5.6.

Q2.5.9. It is likely that quantum theory is the correct way to describe meridians and qi. Quantum theory is, however, too abstract for ordinary clinicians and patients. Can you give a simple analogy?

We could draw an analogy from the relations between a person and human society. A person has a multi-leveled structure, as does society. The analogy between their basic elements is as follows:

Person	Society
Charges	Sex
Negative charge	Female
Positive charge	Male
Electric forces	Sexual attractive force
Water molecules	Family
Water clusters	A group of families
Stable water clusters	A group of organized families

In the multi-leveled structure of a human body, the smallest layer consists of positive and negative charges and the electric forces among them. In multi-leveled human society, the smallest units are individuals. There is great sexual attractive force between the two sexes, male and female, just as there is attractive force between positive and negative charges.

The human body consists mainly of water. The rest is muscle, bones and tissues. Society consists mainly of individuals. There are highways, buildings, railway road, telephone wires, etc., that are part of human society.

When water clusters are aligned as in stable water clusters with a permanent dipole moment, it is analogous to a group of families that are organized into an organization or an association with a purpose.

The human blood circulatory system distributes and circulates nutrients, hormones, cytokines, white blood cells, etc. This is analogous to how a highway system distributes and circulates food, goods, mail and policemen, for example.

The human immune system fights off antigens. For society, this is analogous to soldiers fighting off foreign troops.

The most sophisticated part of human society is the central government. The most sophisticated part of a person is his brain. The central government issues orders to be obeyed by its citizens. The brain issues orders to parts of the body, e.g., telling legs and hands to move. Networks of human relations, such as political parties, are analogous to meridians. Political parties do not run central government, but they set the agenda for how government must run the society. Meridians do not issue executive orders, but they are where the ultimate decisions are made.

Qi circulates through meridians. Newsletters, publications, meetings and conversations circulate information among members of associations. Normally qi circulates around the body regularly in twenty-four intervals. A professional association has regular meetings and newsletters or magazines to circulate vital information among its members.

In Summary

We've discussed here how quantum energy dynamics play a fundamental role in human life, and how acupuncture is used at the quantum level to improve human life.

Quantum fields are described by oscillations, whether in an electro-magnetic field, in air, liquid or in a solid. Oscillations in the meridians cause qions that consist of, at least, phonons and photons, to go out to all the other organs and systems of the body, through the meridian channels. The meridian system coordinates all other physical systems to make the body function as a whole. In the brain are oscillations of brain waves. These form what we know as the human mind. The intensity of oscillations in the meridians and in the human brain determines the level of activity of the human organism. There must be a suitable intensity of qi circulating along the meridians for normal functioning of the body, and there must be brain oscillations of suitable intensity to hold the mind in its various levels of activity, from resting, daily activity, to spiritual states.

Acupuncture has been shown to be very effective in regulating the oscillations of meridians and consequent flow of qi in the physical body. In the following chapters we will discuss specific instances of how acupuncture is used to remedy various illnesses, and why it is able to cure; how to use qi to maintain good health; and how qigong masters may emit qi to heal others directly. In the last chapter, we present the mathematical formulations of how quantum theory works in application to meridians and acupuncture.

Chapter 3

Acupuncture – The Meridian System as the Most Fundamental

Meridians and Survival

The most basic function of a human being is to survive, to sustain ones life. Every cell, tissue, organ and system must work together to maximize an individual's chance of surviving. Meridians, being most likely the first system to evolve in living beings, must play a fundamental role in survival.

Q3.1.1. What possible role do meridians have in ensuring survival?

Meridians are most likely the first system to evolve in living organisms, followed by other systems. The circulation of qi in meridians is probably too slow for the purpose of transmitting warning signals to the organism, so the nervous system developed as a faster system to transmit specific kinds of electrical signals. And since meridians are probably too slow to circulate a significant amount of materials to different parts of the body, the blood circulatory system developed to transmit nutrients quickly. The role of the meridian system, as the most fundamental, is to regulate and balance different parts of the nervous system, the blood circulatory system, and, in fact, all of the systems of the body. Qi and its quanta, qions, are the media that carry signals through various meridians to coordinate functions of different organs.

Q3.1.2. How do different systems of the human body help survival?

The most basic human instinct is to survive, and the second most basic is to ensure the survival of the species. For survival we need material and non-material things. Matter is material; information is non-material. The nervous system, the brain, and the immune system are information systems. Let us say you accidentally put your finger in the fire. The peripheral nerve in the finger immediately sends information: a warning signal to the brain. The brain feels the pain, and sends a signal back to the finger to withdraw.

If bacteria from outside comes into the body and launches an attack upon internal organs, the immune system responds by sending signals causing cytokines and natural killer cells in the blood stream to go to work destroying the bacteria.

The matter required by the human body is food. The systems evolved for handling the food matter via digestion and excretion are the gastrointestinal and the urinary systems.

Q3.1.3. How does acupuncture help?

When a person is sick, probably one of his systems is not functioning properly. The sick system sends a signal to the meridians and affects the circulation of qi, stimulating or blocking the circulation of qi. Acupuncture restores the normal circulation of qi and then enables the meridian system to function properly.

In the figure below is a simple representation of stable water clusters.

Think of unaligned water clusters, each with a negative and positive charge, lying in disarray on a blanket.

Then shake the blanket, and the clusters align.

On the left, above, the water clusters in the meridian are not aligned, and qi, as normal oscillations of water clusters, cannot proceed smoothly. This results in a person getting sick. Acupuncture stimulates more natural oscillations, as in shaking a blanket. This causes the water clusters to align themselves, as shown on the right, above. Then qi will proceed without impedance, as propagating waves along the meridian. This results in the person getting well.

There is much scientific and clinical evidence that acupuncture positively influences the normal function of various systems: the reproductive system, nerve system, gastrointestinal system, urinary system, etc. (See Tables 3.1.1-3.1.5 for details.) Acupuncture not only helps one system come back to normal. It often helps more than one system to function properly at the same time.

When acupuncture is used to improve the immune system, both the brain and the blood change. Magnetic imaging shows that the gray matter of the brain is changed by acupuncture – the cytokines and Nk cells are affected.

Acupuncture affects the peripheral nerves. It can cause injured nerves to recover. In experiments on rats, deliberately severed nerves grow back again during acupuncture treatments.

Highlights of Table 3.1.1. The effect of acupuncture on the immune system

- Maps of magnetic resonance imaging of the brain reveals marked signal

decreases bilaterally in multiple limbic and deep gray structures, which causes immune modulation.
- Acupuncture showed significant immune modulating effects on patients with allergic asthma by measuring their peripheral blood parameters: eosinophils, lymphocytes, subpopulations, cytokins, and in-vitro lymphocyte proliferation rate.
- Treatment at ST36 and LI4 on patients with painful disorders increased beta-endorphin, CD3, CD4, CD8, NK cells and monocyte phyagocytosis.

Highlights of Table 3.1.2. Effect of acupuncture on nervous system

- Rats receiving acupuncture regenerated 10mm gaps of the sciatic nerve with more mature ultrastructural nerve organization that had higher numbers of axon density and blood vessel area.
- Electro-acupuncture improved functional rehabilitation of injured peripheral nerves of 60 Wistar rats.
- Stimulation at LI4 resulted in a significant increase in sympathetic and parasympathetic activity and a significant decrease in heart rate for healthy volunteers.
- Capsaicin-sensitive thin afferent fibers mediated by receptors were activated by acupuncture to participate in inhibition of jaw opening reflex of rats.
- Electro-acupuncture increased pain threshold and caused a transient increase in muscle sympathetic nerve activity.

Highlights of Table 3.1.3. Effect of acupuncture on reproductive system

- EA improved anovulation in women with polycystic ovary syndrome, mostly likely through the normalization of hypothalamic-pituitary-ovarian axis.
- 15% of patients experienced an improvement of quality of erection, and 31% reported an increase in their sexual activities under EA for four weeks.
- EA may activate production of body estrogen, measured by radioimmunoassay, RNA dot blot, monoclonal antibody immuno-histochemistry, and computer image analysis.
- Testosterone and dihydrotesterone rose markedly after EA at ST36.
- EA at BL67 corrected malposition of fetus in women.
- Acupuncture normalized hormonal status in women with ovarian hypofunction.

Highlights of Table 3.1.4. Effect of acupuncture on gastrointestinal system

- The frequency of bowel movement increased from 1.4/week to about 5.0/week for children with chronic constipation.
- 12 out of 13 cases showed recovery of normal peristalsis within 72 hours of operation.

- Acupuncture seems to be effective in the treatment of irritable bowel syndrome.
- Intestinal peristalsis was accelerated significantly by acupuncture at the abdomen.
- Gastro-colonic disorder was introduced into rabbits by injection of erythromycine (EM). EA shortened the duration and latency of the effect of EM.
- It decreased the frequency, amplitude and variation coefficient of gastro-colonic electric activity.
- Acupuncture restored motor-evacuatory function of stomach and bladder for 220 patients with purulent peritonitis.

Highlights of Table 3.1.5. Effect of acupuncture on urinary system

- Acupuncture at BL33 controlled incontinence urge, reduced uninhibited contraction, and increased maximum bladder capacity.
- Renal blood flow (RBF) was decreased by EA in normal and induced renal ischemia rabbits. After renal neurotomy, RBF was increased by EA. This suggested EA affected both renal and body fluid.
- Inhibition of the rhythmic micturation contraction on urinary bladder by acupuncture-like stimulation of the perineal area as a reflex response characterized by segmental organization.

Q3.1.4. Can acupuncture help babies in utero?

Yes, during pregnancy if the position of the fetus is not correct, acupuncture can move the fetus to the correct position. It can even help to speed up the birth of the baby. Furthermore, before the conception of the baby, acupuncture can restore the function of the male and female sexual organs. Acupuncture probably affects the sexual organ via the hypothalamus and pituitary glands.

Q3.1.5. Human physical systems consist of many parts. How does acupuncture influence a system?

Acupuncture influences all levels of structure in the human physical system, from the largest to the smallest. Inserting needles at the acupoints of, say, the multi-leveled structure of the gastrointestinal system affects its organs – the jejunum and ileum; its tissues – the pyloric membrane and gastro duodenal mucosal lesion; as well as its molecules and electric fields, e.g., sero-enzymes and beta-endorphins. We can monitor the body's electrical activities by measuring the frequency and amplitude of its electric field.

Q3.1.6. Can you give another example besides the gastrointestinal system where acupuncture influences all levels of a system?

The urinary system is another example of a multi-leveled system, where every level of structure is influenced by acupuncture. The urine of a hyperactive bladder, the flow of renal blood, and molecules like aldosterone rennin or natriuretic peptide, are all affected by acupuncture.

Q3.1.7. So far you seem to imply meridians are all pervasive. The current dominant view in scientific circles is that the brain is the dominant human organ. What is the relationship between meridians and the brain?

Ancient Chinese literature talks very little about the brain, so we are not sure. But we believe an analogy to human society might be useful. Meridians are like a network of human relations. The nervous system is like a system of telephone lines, and blood vessels are like highways. The brain, then, is like the central government. In daily life, human beings work in a network of relations among relatives, friends and business associates. We talk face to face and make exchanges of various kinds. We telephone one another for speedy communication. These face-to-face exchanges might be analogous to qi in a human body. Most of the time the central government is far away and seems to be irrelevant.

The central government plays an executive role, ideally, executing decisions on behalf of the entire state or nation, reflecting the greater will and good of the nation. This can be seen as analogous to the brain, which executes the decisions arrived at and impelled by the requirements of the whole person, physical, mental and spiritual, coordinated by the meridians.

Most parts of the human body function through a network of meridians with qi circulating in them. For quick delivery of nutrients, or for moving wastes, the body uses the blood vessels. When immediate responses are required, such as in an emergency, the nervous system is used. Telephone lines grew out of the need for faster human communications. They improve human relations, but they are no substitute for them. Without a network of human relations, no human society can function or survive. The nervous system grew out of the need for faster communication among different parts of human body. It is no substitute for the meridian system. Without meridians to circulate qi, there is no life.

From a research point of view it would be most interesting to study the relation between the brain and the meridians that cover the head and are closest to the brain, which are the twelve yang meridians and the Du meridian. There are some research studies already that show acupuncture affects the function of the brain. It would be interesting to see how the brain affects the meridians.

Q3.1.8. Traditional Chinese medicine theory does not use the same language as the present day science of anatomy. The internal organs, called Zangfu, (five Zang and six fu) bear the same name but are not the same as those referred to in anatomy. Does acupuncture affect the internal organs, as we know them in modern anatomy?

Yes, acupuncture is guided by the theory of Zangfu, but nevertheless it has an effect on present internal organs. (Shown in Tables 3.1.6-9.)

Highlights of Table 3.1.6. Effect of acupuncture on the liver

- Rats with acute liver damage had higher levels of serum glutamate-oxalate-transaminase and serum glutamate-pyruvate-transaminase. After acupuncture they were significantly reduced.
- Rats with liver damage were treated with acupuncture. Their biochemical and morphological parameters of liver injury were significantly reduced.

Highlights of Table 3.1.7. Effect of acupuncture on spleen

- EA could prevent the decrease of lymphocyte proliferation response of spleen in rats induced by morphine.
- EA at ST36 at 1 Hz enhanced splenic natural killer cytotoxicity, increased interleukin (IL)-2 and interferon (IFN) -gramma.
- IL-2 production was significantly reduced in rats with injured spleen. EA increased the induction of IL-2 afterwards. It implied EA could improve immunosuppression induced by trauma stress.

Highlights of Table 3.1.8. Effect of acupuncture on stomach

- The protective effect of EA on injury of gastric mucosa in rats was due mainly to changes in nitric oxide.
- EA was able to protect stress rats from stress-induced peptic ulcer, probably by enhancing the gastric mucosal barrier, stabilizing gastric mast cells, and inhibiting the gastrin levels in gastric mucous.
- EA abolished consequences of pain-induced acute emotional stress, resulting in significant reduction of gastric erosions.
- EA significantly increased the percentage of slow waves, as recorded by surface electrogastrography, and hence may be an option for treatment of gastric dysrhythmia.

Highlights of Table 3.1.9. Effect of acupuncture on lung

- Laser acupuncture improved bronchial potency, enhanced bronchial sensitivity to sympathomimetics, and reduced systolic pressure in the pulmonary artery for 111 patients with obstructive chronic lung diseases.
- Laser acupuncture was found to improve bronchial permeability, enhance oxygenation of arterial blood, and normalize sleep and appetite.
- Acupuncture helped 12 patients with chronic obstructive pulmonary diseases in terms of subjective scores of breathlessness and six minute walking distances.

Getting well – what acupuncture cures and how it cures

Q3.2.1. People come to acupuncture because they are sick. What can modern science offer beyond what traditional Chinese medical literature offers?

Classical medical literature offers well-tried prescriptions for treating various kinds of diseases. I meet newly graduated acupuncturists who have memorized all the procedures in the textbook. They are amazed at the results they get by just following the textbook in their treatment of patients.

At the present time, Western medical science offers something quite different. It is analytic. It dissects things into smaller and smaller components, focusing its attention on one point in one study. It organizes evidence systematically, compares success with failure, and controls the conditions where diseases are treated. Conclusions are based on statistical analysis. Traditional Western medicine is not holistic, but does offer insights into the mechanism of acupuncture, based on the findings of physics, chemistry and mathematics.

There are currently two kinds of scientific studies on acupuncture: scientific experiments with animals and clinical studies with patients.

Q3.2.2. Why do we do experiments with animals?

It is universally considered unethical to do experiments on humans. We cannot deliberately induce diseases in humans. Many consider it ethical to conduct experiments in animal studies. And there are many practical reasons for doing so. In animal studies, we are in control of the conditions of the diseases. When a disease is cured, we can be certain that it is not a placebo effect. We can insert devices inside an animal without worrying about long-term effects. More data can be collected and crosschecked.

Furthermore, the cost of animal studies is much less. The number of subjects can be larger. The time to carry out the tests is also shorter. The results are reproducible. They carry an objectivity that is untainted by human psychology. The only weakness is that rats are rats and rabbits are rabbits. They are different from humans in many ways. The ultimate test of a successful treatment is still one done with human beings. The results from human trials are less objective, but may be more relevant to curing patients.

Q3.2.3. Cardio-vascular diseases are the number-one killer in industrial countries. What kinds of effects does acupuncture have on cardiac disorders?

Most animal studies and clinical studies focus on using the acupoint Neiguan PC6 of the pericardium meridian. PC6 is located about 5-6 cm above the transverse crease of the wrist. Acupuncture at PC6 affects all levels of the human body, from the smallest particles of electrons and photons, to the larger molecules and cells, and up to the level of organs and systems. (For details, see Fig. 3.2.3 and Table 3.2.1.)

The approach of Western medicine is to focus on one aspect of a disorder, and then to develop one particular treatment, such as a drug or surgical procedure. The treatments are specifically designed to cure that disorder without affecting anything else. For instance, heart by-pass surgery is designed to accomplish just one thing: the by-pass, hopefully without affecting the nervous system, the heart itself, or the spinal cord.

The approach of acupuncture is different. It aims, through affecting the meridian system, to affect the heart, the nervous system, the spinal cord and the blood circulatory system, thereby balancing the various body parts so that they coordinate to function together normally.

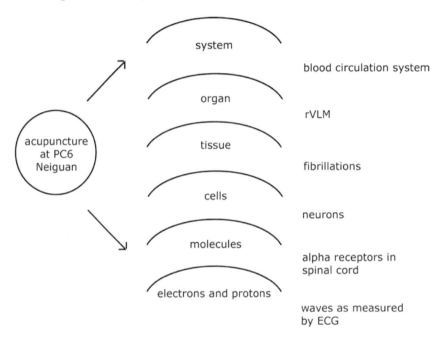

Fig. 3.2.3. The effect of acupuncture at PC6 for cardiac disorders is on multiple layers of the body. The multiple layers are from large to small: system, organ, tissue, all the way down to electrons and photons. Details are given in Tables.

Highlights of Table 3.2.1. Effect of acupuncture on cardiac disorders

- For open chest dogs, EA at PC6 improved mean arterial pressure, end-diastolic volume, heart rate, stroke volume, cardiac output, end-systolic pressure.
- EA could relieve arteriolospasm, inhibit extreme dilation of blood capillaries, modulate imbalance of microvasomotion of coronary artery, improve myocardial blood supply, and improve normalization of electrical activities for rabbits with acute myocardial ischemia (AMI).
- Rostral ventrolateranal medulla (rVLM) participated in the regulatory effect of EA on the heart.

- EA at PC6 could significantly accelerate the recovery of ST-segment and T-wave of electrocardiogram (ECG) induced by AMI.
- Acupuncture at PC6 influenced the superficial electrical resistance of acupoint PC6 and increased the heart rate. It is suggested that PC6 and heart may be connected through a short reflex.
- Acupuncture regulated and improved heart rate variability in 20 coronary heart disease patients.
- 8 indexes of electrocardiogram, cardiophonogram and rheocardiogram all changed after acupuncture at PC6, PC4, PC3 and PC2 for 100 coronary heart patients.
- During acupuncture the number of angina attacks per week was reduced from 10.6 to 6.1.

Q 3.2.4. Books on acupuncture list hundreds of sicknesses it cures. Can you give just a few examples of sicknesses acupuncture cures, which Western medicine finds most difficult to deal with?

Some of the most difficult diseases facing Western medicine are cancer, arthritis, diabetes, and diseases associated with the autoimmune system, such as allergies (Table 3.2.2-5). Acupuncture has curing effects on all of these, to varying degrees.

Highlights of Table 3.2.2. Effect of acupuncture on arthritis

- Acupuncture may enhance the release of a peripheral opiate-like substance, leading to a more potent analgesic effect in the inflamed area of adjuvant arthritis rats.
- During acupuncture analgesia on adjuvant arthritis rats, the delta waves were decreased and beta waves were increased. They were measured from cortical and hippocampal EEG.
- Significant improvement was obtained from 32 patients with osteoarthritis of the hip.
- Acupuncture at SP9, SP10, ST34 and ST36 significantly reduces symptoms in 44 patients with advanced osteoarthritis of the knee.
- The natural killer (NK) cells activity and IL-2 values in patients with rheumatoid arthritis were found to be lower than those of normal persons. Both of these increased after acupuncture treatment

Highlights of Table 3.2.3. Effect of acupuncture on allergies

- Acupuncture exerted immunomodulatory effect on asthma. CD3+, CD4+ cells increased significantly. IL-10 decreased, IL-6 increased, lymphocyte proliferation rate increased, and the number of eosinophils decreased.
- Difference between lymphocyte transformation, acidophil cell count, IgA, IgG, E-rosette formative rate between treatment group and control group were significant.

- By using peak flow variability, spirometric analysis and markers in blood and sputum, a positive effect was found on 66 patients with mild bronchial asthma.
- Ig A in saliva, nasal secretions, and IgE in sera were significantly decreased from acupuncture on asthmatic patients. This suggested acupuncture could inhibit allergic asthma attacks.

Highlights of Table 3.2.4. Effect of acupuncture on cancer patients

- Experiments with implanted mammary cancer on rats found that there were differences in pathological section, adenoid structure, lymphocytic infiltration and tumor volume between treatment group and control group. Acupuncture probably inhibits growth of mammary cancer and reduces its malignancy.
- Acupuncture had a dramatic effect on xerostomia, dysphagia and articulation on cancer patients suffering from xerostomia. Release of neuropeptides that stimulate the salivary glands and increase blood flow is a possible explanation.
- Acupuncture enhanced the cellular immunity of patients with malignant tumors: % of CD3+, CD4+, and ratios of CD4+/CD8+ and the level of beta-EP were increased, and the level of SI-2R was decreased.
- Acupuncture was done on breast and uterine cancer patients who had suffered skin injuries from radiation therapy. Radionuclide and rheographic studies as well as evaluation of hemostatic function showed acupuncture to be effective for edema and pain. It also improved lymph flow, rheovasographic indexes, and normalized stasis.

Highlights of Table 3.2.5. Effect of acupuncture on diabetes mellitus

- EA at EX-B3 and ST36 significantly lowered blood glucose and release of pancreatic glucagons.
- Stimulation at ST36 increased cell proliferation and neuropeptides Y levels.
- For 46 diabetic patients, 77% showed significant improvement in their primary and secondary symptoms.
- By using rheovasography, thermograph, and ultrasound dopplerography it was found that acupuncture improved elastotonic properties of arteries of average caliber, and enhanced blood outflow and regulation of lower limb vascular peripheral resistance for 55 insulin-dependent patients.
- Laser acupuncture removed pain syndrome, improved peripheral circulation and function of lower extremities and obliterated athero-sclerosis of the legs.

Q 3.2.5. How does acupuncture cure arthritis?

When we use acupuncture to treat arthritis in the knee or any other place, all levels of the body are affected (Fig. 3.2.5). The treatment affects the spinal cord, reduces swelling of the soft tissue, suppresses the activities of pain sensitive neurons, increases the activity of natural killer cells, increases the value of interleukin IL2, and changes the pattern of brain waves.

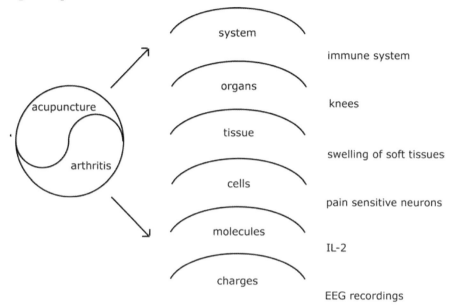

Fig. 3.2.5. Acupuncture can influence all layers of the human body to relieve the various symptoms of arthritis.

Acupuncture pushes various parts of the body to a normal state. If there is too much activity, it reduces it. If there is too little activity, it increases it. The effect is bi-directional. The effect is achieved without injecting any chemicals or inserting any artificial devices. It does this by stimulating the natural mechanisms of the body. Acupuncture treatments have no side effects, because nothing foreign has been introduced permanently into the body.

Q3.2.6. There are estimates that half of the Chinese from Hong Kong who immigrate to the United States will become allergic to something after several years of sensitizing. It could be to oak tree pollen, dust from a steel mill, or simply some milk product. Can acupuncture help?

A study by the University of Vienna indicated that 70% of patients reported significant improvement in allergic conditions after ten weeks of treatment. Another study showed that patients who were under cortisone treatment could reduce the amount of cortisone they took and still improve their condition when they were under acupuncture treatment.

Acupuncture achieved such results by influencing the cellular and molecular components of the human body, such as eosinophils, lymphocytes, cytokins and immunoglobulins (Fig 3.2.6 and Table 3.2.3).

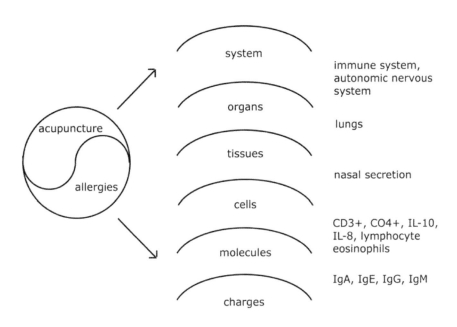

Fig. 3.2.6. Acupuncture can influence all layers of the human body to relieve symptoms of allergies.

Q3.2.7. Many people worry that when they have something abnormal happening to their body that does not go away under normal treatment, they might have cancer. Could acupuncture help also?

Animal studies suggest that electro-acupuncture may reduce the incidence of breast carcinoma. However, most recent scientific studies focus on reducing side effects in cancer patients after their radiotherapy and chemotherapy treatments. Acupuncture can relieve the symptoms of cancer treatment, such as dry mouth, vomiting, breathlessness and flashes. Internally it relieves pain and bolsters the immune system. The red blood cells and white blood cells are affected by acupuncture to make patient feel better (Table 3.2.4).

Q3.2.8. Newspapers in the US continuously report the increase of persons among all age groups who suffer from diabetes. Diabetes becomes more and more prevalent as the society gets more and more affluent. It is caused primarily by dysfunction of the pancreas. What can acupuncture do for diabetic patients?

Scientific studies and clinical tests in international research centers in the last ten years have shown that acupuncture can help diabetic patients in the following ways:
- Lower blood glucose content
- Lower the release of pancreatic glucagons
- Attenuate symptoms of polyphagia (the urge to eat too much), polydipsia (excessive thirst), polyuria (excessive passage of urine)
- Prevent slowing of motor nerve conduction
- Improve microcirculation, and myocardial contractility
- Enhance blood outflow and regulate vascular peripheral resistance
- Have an antiatherogenic, antioxidant, and immunomodulating effects
- Obliterate atherosclerosis of legs
- Induce secretion of endogenous beta-endorphin
- Elevate the lowered pain threshold
- Increase cell proliferation and neuropeptide Y level

For details see Table 3.2.5 and Fig 3.2.11. We show infrared pictures of one diabetic patient who had had laser surgery on the eyes without much success. Then he had an acupuncture treatment. The effect of the acupuncture is dramatically illustrated by the change in the patient's color pattern (Pictures 3.2.1, 3.2.1a).

Highlights of Table 3.2.5. Effect of acupuncture on diabetes mellitus
release
- EA at EX-B3 and ST36 significantly lowered blood glucose and of pancreatic glucagons.
- Stimulation at ST36 increased cell proliferation and neuropeptides Y levels.
- For 46 diabetic patients, 77% showed significant improvement in their primary and secondary symptoms.
- By using rheovasography, thermograph, and ultrasound dopplerography it was found that acupuncture improved elastotonic properties of arteries of average caliber, and enhanced blood outflow and regulation of lower limb vascular peripheral resistance for 55 insulin-dependent patients.
- Laser acupuncture removed pain syndrome, improved peripheral circulation and function of lower extremities and obliterated atherosclerosis of the legs.

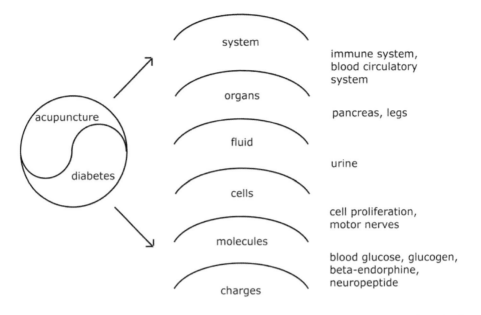

Fig. 3.2.8. The effect of acupuncture on diabetic patients is on all levels of the body.

Q3.2.9. It is well known in the West that acupuncture is especially good for relieving pain. How does one know that there is relief of pain in patients?

Pain is a complex phenomenon and has many causes. At the moment most assessment of pain comes from patients themselves. This is done through filling out various standard questionnaires, assessing numbers on a visual analog scale, or assessing from a disability index. The analgesic effect of acupuncture, however, can be measured by the amount of beta-endorphin released. (For results of various studies of headache, neck pain, shoulder pain and lower back pain, see Table 3.2.6-9).

Infrared imaging offers an objective method for assessing the relief of pain by measuring the decrease of temperature of hot spots, or increase of cold spots on the surface of skin, resulting from acupuncture (Picture 3.2.2-7). If a needle is put in the wrong spot, there is no change. Or sometimes the condition is worsened: there is an increase of temperature in hot spots, and a decrease of temperature in cold spots. When that happens, the practitioner can change the needle position until the desired result is achieved. According to the limited experience of the author, all the acupuncturists working with him so far can achieve positive effects, as certified by infrared imaging, almost always on each treatment, when circumstances allow full cooperation among the patient, the acupuncturist and the author.

Highlights of Table 3.2.6. Effect of acupuncture on headache

- A Finnish survey of 348 patients showed significant relief of pain in myofascial syndromes.

- EA was effective in treating 700 patients with cranio-facial pain.

Highlights of Table 3.2.7. Effect of acupuncture on neck pain

- Acupuncture showed greater improvement than massage for 117 patients with chronic neck pain.
- Randomized single-blind test on 46 patients with chronic myofascial neck pain were relieved with relevant acupuncture with heat.

Highlights of Table 3.2.8. Effect of acupuncture on shoulder pain.

- Deep acupuncture is shown to be better than superficial acupuncture for treating 44 patients with shoulder myofascial pain.
- Acupuncture improved 52 sportsmen with rotator cuff tendonitis, as shown by a modified Constant-Murley score.
- EA plus nerve block by xylocaine were shown to successfully treat 150 patients with frozen shoulder.

Highlights of Table 3.2.9. Effect of acupuncture on low back pain

- Acupuncture was superior to control regarding pain intensity, pain disability, and psychological distress for 131 patients suffering with chronic back pain for many years.
- Intensity of low back pain dropped from 59 to 19 mm and intensity of radicular pain from 64 to 12 mm, as assessed by visual analog scale.
- Acupuncture massage showed beneficial effects for disability and pain compared with Swedish massage.
- Acupuncture was shown to be better than physiotherapy for treating pregnant women with low back and pelvic pain, as shown in pain scores and disability rating index.
- EA with 2 Hz was better than 80 Hz in treating 40 patients with low back pain.

Q3.2.10. From your infrared imaging study, what are the most significant new findings about pain?

There are several things. First, it is quite common for infrared pictures to reveal that the pain the patient complains about is not the most serious one. The patient may complain about pain and numbness in the hands, but the hottest spot turns out to be at the back of his neck. Without curing the hot region of neck, concentrating on the pain in the hands may not cure the root cause of the pain.

Second, the infrared picture often reveals an old injury that has not been cured. I took a picture of the back of my nephew once. He had complained about the right side of his back. I discovered that there was a hot spot on the left shoulder. He said he had had a car accident when he was twelve. He felt soreness there when it was humid and raining. On normal days he felt nothing. It was quite logical to expect such a spot to become a serious problem when he became old or weak.

Third, there are two kinds of pain: pain from a cold region and pain from a hot region. Pain from a cold region is generally an old problem, and will take longer to cure. In our experience, moxibustion seems to be particularly good for pain in a cold region.

I am sure more new things will be found out in the future when more and more cases are compiled.

Symmetry in human beings

Q3.3.1. It is well appreciated in art that symmetry in objects evokes a sense of beauty in the observer. Does symmetry have anything to do with science, especially medical science?

Symmetry does not play an important role in physics until the 20th century. Some would say the first 50 years of the 20th century of physics were dominated by symmetry. The event of quantum theory elevates symmetry to a fundamental status. Also Einstein's special and general relativity theories incorporate symmetry as one of its foundations. The second half of the 20th century in physics was dominated by symmetry breaking, starting with the discovery of left-right symmetry violation by C. N. Yang and T. D. Lee.

There are three fundamental symmetries in quantum field theory: the left-right (space) symmetry, the symmetry with respect to the change of charges, and symmetry in time. Surprisingly, these symmetries and their breaking seem also to hold an important role in medicine.

Q3.3.2. Can you give us some examples of left-right symmetry and asymmetry in Oriental medicine?

Meridians obey left-right symmetry. On the left foot there is a stomach meridian. Opposite, in the right foot, there is again a stomach meridian. But DNA molecules have left-handed spirals only. No DNA molecule has a right-handed spiral. Hence DNA molecules are left-right asymmetric. When a person is well, qi circulates in a left-right symmetrical way. When the left-right symmetry in qi circulation along meridians is broken, people become sick. When qi is blocked, people will feel pain. Blocked qi sometimes shows up as left-right asymmetry in the temperature distribution of the skin surface. With acupuncture treatment, left-right symmetry is restored, and pain is relieved. Thermal distribution of the person becomes left-right symmetrical again (Table 3.3.1 and Pictures 3.3.1-2). A person who has suffered from stroke often has left-right asymmetry in thermal images of the body. Acupuncture can restore the left-right symmetry of the body, and the person is cured.

Q3.3.3. Since Oriental medicine talks about yin and yang, or basically negative and positive charges, is there any symmetry in dealing with charge?

Yes, charge symmetry occupies an important role in quantum theory. Let us give a simple example of two arbitrarily distributed positively and negatively charged clusters. They attract each other by electric force. If we change the positively charged cluster into a negatively charged cluster with the same distribution, then the attractive force becomes a repulsive force with equal magnitude, but opposite in sign. It does not matter what the distribution of charges is. Neutral objects such as photons, or neutral qions, do not change into different particles under charge symmetry.

Q3.3.4. What does charge symmetry do to a living system?

We are not so sure yet. But from animal studies we know that the growth point of a tissue is negatively charged. When it becomes positively charged, the growth stops. In treatment with electro-acupuncture it is, then, important to have the right polarity for the right symptom.

Q3.3.5. Besides symmetries in space and charge, is there symmetry in time?

Yes, there is symmetry in time. Time is the dynamic variable in any elementary motion. Its symmetry has a profound effect on motion. When we play billiards, we see billiard balls bouncing off one another. If we take a videotape of the motion of the balls and then play it backward in time, we cannot tell whether the video is going backward or forward. They are equally plausible motions. The fundamental reason is that in classical mechanics Newton's law does not change if we change the sign in time. We say that the equation of motion is invariant under time reversal.

Similarly if we videotape the vibration of a piano wire, we could not tell the difference if we ran the tape forward or backward in time. We think that oscillations in the meridians are the same. These oscillations are time reversal invariant.

During treatment by moxibustion, infrared radiation goes from acupoints, through meridians, to the intended internal organs. By time reversal invariance, infrared radiation can also go from an internal organ, through a meridian, to acupoints. So if one of the internal organs is not well and has a high temperature, the signal of high temperature will be carried by infrared radiation from this internal organ through meridians to acupoints. Hence temperature variation of acupoints on meridians could serve as a diagnostic tool for dysfunction of internal organs.

Similarly, treatment of electro-acupuncture uses AC signals to reach the internal organs. By time reversal invariance, it is also possible to measure the AC signals on the acupoint to determine the state of affairs of internal organs.

Time reversal invariance connects the treatment process with diagnosis. A living being is never as simple as inanimate elementary objects like billiard balls. We expect there will be deviation from exact invariance, and we expect surprises.

How do you know you are sick? – diagnosis

Q3.4.1. What are the traditional ways in Oriental medicine to find out how sick a person is? How can we modernize them?

There are four ways for a traditional Oriental doctor to find out how sick a person is: to observe, to ask, to listen and to feel. Asking and listening are part of the doctor/patient relationship. These should remain sacrosanct. We can modernize the other two ways, to observe and to feel, by using scientific equipment.

Q3.4.2. What kind of scientific equipment would you use to simulate and to improve on "observing"?

To observe, or to see, is to use light. Light is composed of photons. Photon detectors of any kind could be used to yield objective data about a person's condition. X-ray machines use photons with very short frequencies to penetrate the body and yield internal information from the scattering of photons inside the body. Magnetic Resonance Imaging (MRI) uses photons with very long wavelengths in the range of radio frequencies. In association with high magnetic fields it can yield information on the distribution and conditions of protons inside the human body. We propose here in the first instance to use photons in the infrared range to detect the temperature distribution on the surface of human body. From deviation from standard temperature at acupoints one can deduce the dysfunction of organs or tissues along the meridian where acupoints are situated.

Q3.4.3. How do we modernize "to feel"?

We can use a device, which could be mechanical or electrical. One may build a pressure-sensitive device made up of transducers, to feel the pulse. It seems rather difficult so far to quantify the palpitation method of a traditional Oriental doctor. But one could build an electrical device that measures electrical conductivity at the acupoints. Such devices have been designed in China, Japan, and Germany, notably, the German Voll meter. These machines are used with a personal computer and automatically display the condition of a person's meridians.

Q3.4.3. Let us come back to infrared imaging. Why is temperature important?

Temperature is a measure of thermal energy. When a person goes to a clinic feeling sick, the first thing a nurse does is measure his temperature orally. This oral temperature, normally 37° C, tells us whether the biochemical reactions inside the body are proceeding at a normal rate. An infrared image, which can be taken in seconds or sometimes in a fraction of a second, will give tens of thousands of temperatures on the surface of the skin. This is an extraordinary amount of information about the rate of biochemical reactions going on inside the body.

Q3.4.4. How do we use the rich source of information contained in a thermal image?

We could proceed empirically and document what kinds of diseases have what kind of pattern of thermal distribution on an infrared image. This is a hard and laborious method. Or we can proceed much faster and more accurately with the guidance of a theory. The theory we have is meridian theory, which connects body phenomena with internal organs. To simplify matters greatly, we can say a meridian is like a fiber optic tube that one can use to look at the temperature of internal organs through acupoints on the meridian.

Q3.4.5. In Q3.3.5 we say that the relation between diagnosis and treatment is a relation connected by time reversal. Can you explain more?

Acupressure is a treatment that applies mechanical pressure on acupoints. The pressure is transmitted by meridians, as a wave, to cure internal organs. The sickness of an internal organ will also be signaled via meridians transmitting pressure waves to special points. A doctor can then use his hands to detect the signal, which appears as a variation of palpitation.

Electro-acupuncture treatment uses electrical signals, which are transmitted by meridians to internal organs. Internal organ sickness will also be reflected as electrical properties at the acupoints.

Moxibustion treatment uses heat, transmitted through the meridian at the acupoint, to internal organs. Sickness of internal organs will also transmit information, as heat, through meridians to acupoints, which heat can be detected by infrared imaging.

Q3.4.6. How does infrared imaging help an acupuncturist determine the source of pain when the patient oftentimes explicitly tells the acupuncturist where the pain is?

We should always listen carefully to what the patient says. There is always truth in it. But our experience has shown that the patient tends to complain about the most recent pain, which may not be the source of the problem. In Picture 3.4.1, we show a woman who complains about her back, but she has two other areas that are more problematic – the neck and the middle back, revealed by infrared imaging. Infrared imaging reveals the hottest spot, which signals the hot pain area, and the coldest spot, which signals the cold pain area. It is quite logical to deduce that the higher the temperature, the more serious the problem. The source of the problem is not necessarily what the patient complains about. Another case is shown in Picture 3.4.2.

Q3.4.7. Can infrared imaging reduce the chance of misdiagnosis?

In Picture 3.4.3 we show a case where a patient avoided back surgery, because infrared imaging revealed that his chief source of pain might be in the head instead of the back.

Q3.4.8. Can you give us an example where high temperature at some acupoints in the limbs may reveal problems with internal organs?

An abnormally high temperature at PC6 might indicate problems with the heart (Picture 3.4.4). Abnormally high temperature at the urinary bladder meridian acupoint, BL1, and the stomach meridian acupoint, ST36, might indicate problems with the bowel movement and stomach.

Getting rid of pain exponentially – evidence from infrared imaging

Q3.5.1. Acupuncture can help people reduce pain quite well. But does acupuncture cure the pain, or merely numb the feeling of pain, the way painkiller drugs do?

There are many sources of physical pain, and hence as many ways to reduce it. Painkillers block the pain signals from reaching the brain, so the patient feels no pain and can function as usual. The source of pain, however, is not treated.

Another method is to find the source of pain and treat that. We can use infrared imaging to locate the hot spots, or sources of pain. This is analogous to taking aerial photographs of a countryside where a fire is raging in order to locate the position of the fire. We can then use acupuncture to extinguish the fire, or pain, as we monitor the decrease in temperature of the troubled area. When the temperature becomes normal again, the patient no longer feels pain. The source of pain is gone and the pain is cured rather than numbed.

Of course, there may be instances where the temperature of the troubled area does not become normal after acupuncture treatment, and then other kinds of treatment are required.

Q3.5.2. How quickly and in what manner does acupuncture cure pain?

Let us consider hot pain in the soft tissues. With one needle we can reduce the temperature. The temperature decreases exponentially over time. (See Table 3.5.1 and Picture 3.5.1.)

We want to decrease the temperature by 36.78%. The time constant runs from 4.3 minutes to 63.58 minutes, with an average of 17.5 minutes, for the 16 cases of pain studied. If we wait for twice as long as an average time constant (=17.5 minutes), which is about 35 minutes in the course of one normal acupuncture session, then we will probably achieve 86.5% of the desired healing effect possible through acupuncture. The longer the acupuncture session lasts, the better the result is, but with a diminishing rate of results.

Q3.5.3. How do you explain this exponential law of the decrease of temperature during an acupuncture session?

There are probably many ways to derive such a result. Let us use a simple model. Let us assume the soft tissue is made up of spherical cells. During injury these cells get deformed. Distress signals are sent to indicate pain. Biochemical reactions are increased to pump energy into the cells so they can repair themselves. Extra heat is generated by the increased biochemical reactions, which heat is detected by infrared imaging devices.

Acupuncture initiates a healing mechanism that oscillates these cells in a random statistical fashion. (For details of mechanism, see next chapter.) During the oscillation of these deformed cells, some, but not all, of them would return to normal spherical shape. The number of cells returning to normal is proportional to the total number of deformed cells in any given moment. When deformed cells return to their spherical shape they no longer need to pump up the biochemical reactions. No extra heat is emitted and the temperature is reduced. This general statistical mechanism, which is intrinsic in any quantum process, would ensure the exponential law is observed. (For mathematical derivation of this law, please see Chapter 5.)

Q3.5.4. Acupuncture claims to use the internal healing mechanism of the human body to cure disease. Hence it is bi-directional. What does this mean? Is there any evidence for this?

Western drugs are designed to work in one direction only. If a drug is designed to reduce temperature in a fever, it will not raise the temperature of the body in the case when the body is cold. A drug that reduces the rate of biochemical reactions will not increase them. A drug that increases the rate of biochemical reactions will not decrease them.

Acupuncture is different. If pain is indicated by hot areas in infrared imaging, acupuncture will reduce the temperature in the hot areas. If the patient with pain has a cold area, as revealed by an infrared image, acupuncture will increase temperature in the cold area.

Moxibustion produces heat. Normally one expects it to heat up the area where it is applied. In the case of pain, moxibustion applied properly will reduce the temperature of the area requiring healing (Picture 3.5.2).

What acupuncture cannot do

Q3.6.1. For the casual reader of an acupuncture book, it looks as though acupuncture can cure any disease. Is there any limit to what acupuncture can treat?

In order to properly evaluate the power of acupuncture it is necessary to know its weakness. For a multi-leveled human body acupuncture is directly effective on the largest system, the meridian system, and on the smallest level, the charges. The effect on other levels of the body, such as tissues, cells, or molecules is indirect. By moving charges around, acupuncture can influence molecular reactions, which in turn affect the cells. When cells are changed, the larger structures in the human body are altered.

Acupuncture's indirect methods in general are slower and gentler than the direct methods employed in Western medicine. Any disease that requires an immediate direct operation or replacement of a certain part of the body is probably not the right place to use acupuncture.

Q3.6.2. Can you give some examples of direct approaches?

- Replacement of a lost limb with an artificial limb.
- Laser surgery to attach a detached retina.
- Blood transfusion when there has been enormous loss of blood from injury.
- Periodical injection of insulin for a diabetic patient who has lost the ability to secrete insulin.

In none of the above examples would it be appropriate to use acupuncture.

Q3.6.3. If acupuncture and meridian theory are limited, is there room for Western medicine?

In the multi-level view of the human body, Oriental medicine works on the level of systems and charges, whereas current Western medicine works mostly on the molecular and cellular levels. Optimal treatment of a particular disease may require a direct or an indirect treatment at a particular level.

This decision must be made by the patient and the doctor. Certainly, there is room for both types of medicine. It is expected that in the future both branches of medicine will merge together under the guidance of fundamental science. In particular, quantum theory is expected to play a more and more important role as medical science develops.

Chapter 4

From Vibration to Vibrant Health

To become healthier

Q4.1.1. If you are not sick and are able to go to work, but want to be healthier, is there anything acupuncture can do to help?

When people are without major diseases, their task is to keep healthy and reduce the chance of getting serious sickness. To be healthy is to be in a balanced state of yin and yang. Yin and yang are general terms that can represent many things. We want a balance of physiological and non-physiological elements. By non-physiological elements we mean the mental and spiritual states. Intangible elements that cannot be considered mental or physiological elements, we include in the spiritual domain.

The balance we want, more specifically, is a balance in an excited state, as defined in the simple model discussed in Chapter 2. Acupuncture can help restore balance in the physiological elements.

Q4.1.2. Can you give an example of a person who is unhealthy and out of balance?

Obesity is probably one of the most common unhealthy states in people living in advanced industrial nations. To be healthy, it is necessary to have a balanced absorption of nutrients and consumption of energy. If a person has an excessive absorption of nutrients and insufficient consumption of energy he will become obese. There may be non-physiological factors, such as social environment or personal relationships, which lead a person to have excessive intake of food. Acupuncture may not help in such circumstances where non-physiological factors are important. (Table 4.1.1)

Q4.1.3. How can acupuncture help to reduce weight?

From the study of rats and humans we understand that acupuncture balances the body through its action on all levels. (See Table 4.1.1 for details.)

- On the human body, acupuncture improves the obesity indices, which include weight; body mass; circumference of the chest, loin, hip, and thigh; and their ratios.

- On the systems level, it improves the vegetative nervous system, the respiratory system and the digestive system.
- On the organ level, it regulates the hypothalamic area (LHA) of the brain.
- On the body fluid level, it regulates saliva secretion.
- On the level of body fat, it improves the lipid indices and reduces the percentage of body fat.
- On the molecular level, it improves energy metabolism indices. It reduces the level of noradrenaline and improves aldosterone and the high-density lipoprotein cholesterol.
- On the atomic level, it improves blood sodium and blood potassium.
- On the electron level, it evokes potential in the hypothalamic ventromedial nucleus (HVM).

The scientific study of weight reduction through acupuncture is only beginning. Nevertheless, from such limited study it is clear that acupuncture reduces weight in a more fundamental, balanced and hopefully long-lasting way.

Highlights of Table 4.1.1. Effect of acupuncture on obesity

- The anti-obesity effect of acupuncture might come from effective regulation of the hypothalamic area, where the level of noradrenaline, serotonin and ATPase were altered.
- Photo-acupuncture lowered levels of blood lipids, glucose, cortisol and triodothyronine on 202 children with simple obesity.
- Combined application of acupuncture, moxibustion and auricular acupuncture very effectively regulated the somatotypic indexes of body weight, circumference of the chest, loin, hip and thigh, the ratio of the loin, sebum thickness and body mass index.
- Acupuncture changed the blood sodium, blood potassium, aldosterone, and mOsm plasma to normal levels. It improved water and salt metabolism by regulation of the nervous system and body fluid.
- Using glass microelectrodes, the ventromedial hypothalamic neuronal activity was found to be excited, indicating satiation formation and preservation.

Q4.1.4. People are living to a more and more advanced age. Is there any way that acupuncture can retard the aging process?

A proper use of the balancing effect of acupuncture is expected to delay the natural aging process of the human body (Table 4.1.2). A study done on female rats implies that acupuncture could do the following:

- Shorten the sexual cycle and increase its frequency.
- Improve memory loss and the decrease of immune responses.
- Increase the frequency of neuronal discharges in locus coeruleus.

- Raise the catecholamine/5-hydroxytryptamine (CA/5-HT) ratio in the hypothalamus to delay the aging process.
- Increase norepinephrine and dopamine contents in the brain.
- Enhance mitogenic activities of splenic lymphocytes.

Highlights of Table 4.1.2. Effect of acupuncture on aging

- Catgut embedding at BL23 shortened sexual cycles, increased frequency of sexual cycle, and slowed down the aging process of the genital system.
- EA at BL 23 elevated the frequency of neuronal discharges in locus coeruleus (LC) and increased the activating rate of LC to neurons in medial preoptic area of the hypothalamus. It probably raised the catecholamine/5-hydroxytryptamin ratio in the hypothalamus so as to delay the aging process
- Specific brain regions were assayed for catecholamine contents using liquid chromatography with an electrochemical detector. The mitogenic activities of splenic lymphocytes was also measured. EA increased norepinephrine and dopamine contents in the brain region and enhanced mitogenic activities of splenic lymphocytes.
- Acupuncture at ST36 increased the telomere levels up to a maximum of 2 times the telomere levels before treatment.

The Secret to vibrant health – internal exercises

Q4.2.1. To stay healthy, exercise is necessary. Why do we need exercise?

A clean and tidy room, whether it is a bedroom or a classroom, if left untended, will become more and more disordered. According to the Second Law of Thermodynamics, every object, if left alone, will devolve to a state of maximum entropy, which means a most disorderly state.

A living organism is highly ordered. If left untended, it will naturally decay into a disordered state. A dead body is in a disordered state, and will disintegrate rapidly. To maintain an ordered state and to overcome the Second Law of Thermodynamics requires extra energy. During exercise, the body accelerates biochemical reactions, and extra energy is injected into the ordered state. Any disordered component can be restored into an ordered state from the availability of this extra energy.

Q4.2.2. How many different kinds of exercises are there?

We can classify exercises into two main kinds: external and internal. External exercises emphasize movement of the body for the purpose of toning and training muscles, skin and bones. Most sports are of this type. Internal exercises are those that emphasize movement of parts of the internal body, which are not visible externally.

Their main purpose is to train the internal organs. Qigong is an internal exercise. Some types of qigong exercises specifically aim at improving a particular organ, such as the stomach, lung, heart or kidney.

The effect of these two kinds of exercises cannot be clearly separated. Playing tennis involves running, which accelerates the heartbeat and increases the lung capacity. The goal is not to strengthen the internal organs, but to win the game. Exercise of internal organs is an unintended benefit.

In most qigong exercises, particular movements of the legs and arms are required. But the aim of these exercises is to strengthen the internal organs, not the arms and legs.

Q4.2.3. Is there an easy method to strengthen our internal organs, such as the kidney?

We could rub the skin above the kidney as in a massage. Or we could do exercises in the kidney meridian. As a daily practice to strengthen the internal organs, we could, for example, use our fingers to apply pressure on PC6 of the pericardium meridian to exercise the heart. Or we could apply pressure on ST36 of stomach meridian and LI4 of the large intestine meridian to strengthen general well being.

Q4.2.4. What is the most fundamental part of internal exercise?

The most fundamental part of internal exercise is the exercise of the meridians. The movement of meridians is qi, or quantum oscillations. So exercises in meridians are meant to increase or decrease the quantum oscillations at will. In qigong exercise there is the stage of the small cycle, where the practitioner can willfully guide qi to circulate around the Du meridian and then the RN meridian. Qi will jump from the Du meridian, across the tongue, to the RN meridian and then continue cycling. The next stage of development for a qigong practitioner is to attain the big cycle, where qi will go through to many more parts of the body via many more meridians.

Q4.2.5. What scientific evidence is there for the effect of qigong exercises?

We do not yet have a thorough scientific study of different types of qigong and their effect on different meridians and different internal organs. In Tables 4.2.1-2 we tabulated some work on the effect of qigong in healthy people, and the effect on patients who practiced some form of qigong.

As listed in Table 4.2.1, the effects are as follows:

- On general well being, qigong can stabilize the cardiovascular system, such as heart rate, respiratory rate and systolic blood pressure. It can induce a wakeful hypo-metabolic physiological state. The person who practices qigong becomes more relaxed, more alert and less anxious.
- On the molecular level, it can lower the stress-related hormone, cortisol.

It can greatly increase plasma growth hormone. The insulin-like growth factor is increased in the young. For the elderly, testosterone is increased significantly. There is a change in cytokine production: IL6 and TNF alpha are increased.
- The effect is bi-directional. It can have either a facilitative or inhibitory effect on the visual cortex, depending on the type of qigong practiced.

Resonance as a mechanism for acupuncture

Q4.3.1. In manual acupuncture when the needle goes into the acupoint of a patient, it is important that the patient feels "de qi." What does "de qi" mean in scientific terms?

The term *de* literally means, "to acquire qi." We regard *de qi* as some kind of resonance effect. The concept of resonance can be illustrated by a child's swing. A swing is like a clock pendulum, which has one unique frequency of oscillation. When the swing is empty and we lift the swing up and let it go down naturally, the swing will oscillate back and forth, up and down, at its natural frequency. If a child sits on the swing and exerts force in synchronization with the natural frequency of the swing, the oscillation of the swing will become larger and larger. If the child does not exert force in synchronization, the swing does not move.

In manual acupuncture, the finger twists and turns the needle, goes forward and backward, in order to achieve resonance with the natural frequency of the acupoint and the meridian. The mechanical movement of the needle will make the meridian oscillate more or less at its natural frequency. If the mechanical movement is incorrect or the position of the needle is wrong, there is no alternation of the oscillations in the meridian, the patient does not feel *de qi*, and the effect of acupuncture is minimal.

Q4.3.2. What, then, are the important physical parameters in "de qi"?

As discussed in the example of a child's swing, the oscillation of the swing consists of a natural oscillating frequency and the amplitude of its swings. If it swings higher, it means the amplitude of its oscillation is large. The amplitude has both an absolute magnitude, which is the size of the swing and the phase. The child must exert force at the right moment to increase the swing. To exert in the right amount means the external force must act in phase with the natural oscillation. If the external force is acted out of phase with the natural oscillation, the oscillation may actually decrease, even though the external force acts with the same frequency.

This is the simplest kind of oscillation in physics, which is the oscillation of a plane wave. The oscillations in the meridians, which we call qions, most likely are much more complicated. However, let us stay with only the simplest case, the case of plane waves. The oscillation of a plane wave consists of its amplitude and its frequency. The amplitude has a magnitude and a phase.

Q4.3.3. There is a traditional saying that if qi does not circulate, then pain occurs. When qi recovers its natural circulation, then the pain disappears. How do you explain that?

A piano wire inside a piano will give the right sound in a normal situation. If some dirt sticks to the wire, the wire will not vibrate very much when a key is hit. The piano is out of tune. If we hit the piano wire hard enough to shake off the dirt, then the wire will vibrate at the right frequency and the piano emits the proper sound.

Similarly, if blockage occurs, and the meridian does not vibrate at normal amplitude at its natural frequency, then we feel pain. When a needle is inserted into the right acupoint to cause additional vibration so as to shake off the blockage, then the meridian vibrates freely and the pain is gone.

Q4.3.4. How does acupuncture cure the internal organs?

Let us consider the broadcasting of a TV station. It emits a certain frequency of electromagnetic waves. Such waves go to all nearby televisions. The television that tunes in the specific frequency of the broadcasting station will receive the signal and show the television program. Other televisions that tune in other frequencies will not amplify the signal and show the program of this particular broadcasting station.

In the analogous situation, the meridian system will carry many quantum oscillations (qions) to different sites of the body. Meridians are like the cable in the cable television system. The internal organs that resonate with the frequency of waves caused by acupuncture will be cured. Other internal organs will not be affected.

Q4.3.5. There are different kinds of movement of the needle in acupuncture. It is often emphasized that the right kind of movement is important to achieve the desired effect. What does this mean?

This would indicate that there is more than one kind of oscillation by the meridian. Different kinds of movement of the needle in manual acupuncture will stimulate different kinds of vibration of the meridian. We already know from electro-acupuncture that different ac frequencies (say 2Hz and 80Hz) will sometimes have opposite results. We may attribute that to the fact that there are at least two natural frequencies associated with the oscillation of the meridian.

For a given frequency there are also two different kinds of acupuncture: one is to cure excess, and the other to compensate deficiency. To reduce excessive oscillation of a meridian is to reduce its amplitude. To increase a weak and deficient oscillation of a meridian is to increase its amplitude.

Q4.3.6. There are these extraordinary points that do not fit into the pattern of meridians. How do you explain them?

Since these extraordinary points have clinical values quite similar to acupoints on the meridians, the structure and function of the extraordinary points and ordinary acupoints must be similar. The extraordinary points are probably also local organizing points. Their origin might have arisen like this: when a single cell in the embryonic stage multiplies, it might first split into two cells, A and B. Cell A becomes one of the special acupoints, and cell B develops into one of the internal organs. Since they come from the same single cell, they probably have the same set of resonance frequencies. Then acupuncture at this special acupoint will have an effect on the internal organs evolved from cell B.

Q4.3.7. What is auricular acupuncture?

Auricular acupuncture is a kind of acupuncture that places a stimulating needle or object only in the ear, and not in any of the conventional acupoints on the meridians. The needle has a curative effect, as demonstrated in recent scientific studies (Table 4.3.1). How does auricular acupuncture work in the resonance scheme?

When we turn the ear upside down, it looks like an embryo in development. So the development of the ear from embryo is probably similar to the development of the extraordinary points, except that now a group of cells splits into two parts, Part A and Part B. Part A develops into the ear, and Part B develops into the rest of the human body. The internal organs, such as the heart or lung, would have their corresponding points in the ear. Acupuncture at these points in the ear will vibrate the same set of natural frequencies as their corresponding internal organs. The signal of these oscillations will be carried by the network of meridians from the ear to the internal organs, or vice versa.

Highlights of Table 4.3.1. Effect of auricular acupuncture

- Auricular electrically stimulated analgesia was found to be effective in reducing the anesthetic requirement by 11%.
- Auricular acupuncture was found to be a valuable alternative therapy for female infertility due to hormone disorders.
- Auricular acupuncture at the heart point reduced blood pressure and improved the effect of left cardiac function with II, III stage of hypertension.
- Auricular acupuncture increased cell birth in dentate gyrus of both appropriately fed and food-deprived adult rats.

Q4.3.8. What is the network of meridians?

So far we have portrayed a simple model of meridians. Actually, besides meridians, there are collaterals that are smaller and subsidiary to the main meridians. There are capillaries that are even smaller than collaterals. This is similar to the blood circulatory system, which has main arteries, ordinary blood vessels, and capillaries. Meridians, collaterals and capillaries are made of similar molecular clusters with permanent electric dipole moments. This network of meridians covers the whole body, and is able to transmit signals from any acupoint to internal organs or tissues that resonate with such signals.

Q4.3.9. What is the role of auricular acupuncture in the theoretical foundation of medical science?

In order to understand the effect of acupuncture in orthodox Western medicine that does accept the existence of meridians, a school of thought has evolved that suggests that acupuncture works through the nervous system. However, there are no major nerves that run through the ears. The fact that various points in the ears could affect the corresponding internal organs of the human body suggests that nerves are not the foundation of acupuncture. See Table 4.3.2 for a summary of evidence that does not favor the nervous system as the system responsible for the working of acupuncture.

Vibrant health

Q4.4.1. What do we mean by "vibrant health"?

Our view of health includes the three levels of existence – body, mind and spirit – the whole person, in relationship to the world around him. A quantum view of health accounts for continual activity and change, reflecting the organism's creative response to environmental challenges.

To be healthy, a person needs to have flexibility in order to stay in balance. Loss of flexibility means loss of health. Flexibility implies a continual adaptation, while preserving or recreating a pattern of balance. This holds at all levels, from the biological system, to mental outlook, to personal relationships, to ones relationship to the world, and ultimately to ones spiritual being.

To have vibrant health means to have perfect balance on all levels of being. Qigong is a system for restoring and maintaining the balance of the whole person. Similarly, in Oriental medicine, the total configuration, the patterns of disharmony, provide the framework for treatment.

In the Western view, human life has three levels of existence: body, mind and spirit. Qigong works on three levels of existence: jin (essence of life), qi, and spirit. A human being has three quantum levels: the physical body, the ground state, or minimal energy state, and the excited state, when the person is fully alive. The excited state splits into the states of mind and spirit. In the tradition of qigong masters the state of spirit is regarded as the highest form of human existence. At the moment we know very little about the state of spirit from scientific studies.

Practicing qigong or similar exercises is a way to get close to the state of vibrant health.

In a state of vibrant health, a person could set and achieve the loftiest goal of his life, given his ability and external constraints. Suppose a person is in jail and loses his legs. He could still write poems or books about his lofty aspirations in life in order to influence the world. This actually happened to many intellectuals over the long history of China.

What can a person with vibrant health do?

Q4.4.2. Qigong masters seem to be good examples of people with vibrant health. Can you tell us more about what a qigong master can do? Is there such a thing as external qi?

Genuine qigong masters can emit external qi, but it can't be seen with the naked eye. External qi is very mysterious to most people. Should one believe it or not? Added to the mystery is the fact that none of the qigong masters I met could explain their external qi in scientifically understandable terms.

So let me explain external qi first in the following way: When we hit the piano wire gently with our finger, we can feel the piano wire oscillating, but we do not hear any sound. The oscillation of the piano wire has not transferred enough energy to the air to form a sound wave. If we hit the piano wire harder with a piano key hammer, the piano wire oscillates, and we also hear the sound coming out from the piano.

Analogously, qi is oscillations of the meridians. If the oscillation gets big enough it will be emitted from the meridians as external qi. One of the strongest evidences for the existence of external qi comes from the effect of external qi on animals (Table 4.4.1). Let me describe my own experiences with the external qi of one qigong master, Master Zhou, who practiced in Los Angeles.

I knew Master Zhou for many years. He was rather short, and always had a smiling face. He was kind and considerate to patients. He treated patients by emitting external qi from his hands without touching the patient's body. While external qi itself is not observable with the naked eye, the effect of Master Zhou's external qi was exceptionally easy to see. He could sprinkle some water on a paper towel, place the wet paper towel on the problem area of the patient, then move his hands over it. Water in the paper towel heat up. One could see white water vapor coming off the paper towel. The patient would to feel the heat. When the heat became unbearable, the patient would cry out and ask Master Zhou to stop. When he took away the paper, one could see the skin of the patient had turned red from the heat.

I brought many skeptics to Master Zhou so they could experience for themselves the pain from the heat of Master Zhou's qi. Every skeptic became a believer after that. This ability of Master Zhou was well known, and was taped and shown on television.

Master Gao, who taught me qigong, could also treat people using his external qi. She once rented a room at California State University in Los Angeles to teach qigong. When I was her student, I also took on the duty of answering incoming telephone calls. A call came in from a man who said his mother was sick and confined to a wheel chair, and that he would like to bring her to see Master Gao. I gave him the address and sometime later they arrived. He told us his mother had been in a wheel chair for a long time. Master Gao stood in front of her while we all stood behind her, watching. She moved her hands without touching her. After a while Master Gao asked the woman to stand up. Indeed, she stood up as she was asked.

There were many other cases where I observed the effect of Master Gao's external qi. Some people would move their body involuntarily, or sweat prodigiously. Different people had different reactions.

Several books have been written on various effects of external qi emitted by qigong masters. See, for example, *Scientific Basis of Qigong*, by W. Xie Hsieh, Beijing University of Science and Technology Press, 1988. We have summarized certain physical properties of external qi emitted by qigong masters in Table 4.4.2, as reported in Hsieh's book. The frequency range of these external qi emissions is from 0.05Hz to 10 Hz. For external qi from qigong masters, the very low frequency, below 1 Hz, is particularly significant. It is interesting to note that these below-1 Hz frequencies are similar in frequency range to those measured in the oscillations of stable water clusters. (See References, Chapter 1, Bibliography for further reading.)

The likely explanation is that the stable water clusters line themselves up in a much more orderly fashion along meridians in qigong masters than those in ordinary people. When qigong masters think about emitting qi, the stable water clusters in their meridians oscillate more and emit more photons and phonons. This is similar to what happens with athletes who think about jumping high, causing the muscles in their legs to contract, enabling them to jump higher.

Highlights of Table 4.4.1. Effect of external qi on animals

- External qi might inhibit N-acetyltransferase and increase serotonin level, as measured in EEG of rabbit and electrical activity of rat pineal gland.
- External qi had inhibitory effects on tumor growth and enhanced effects on anti-tumor immunologic functions of tumor host. With chemotherapy it increased the anti-tumor efficacy.

Highlights of Table 4.4.2. Physical properties of external qi by qigong masters -- photon and phonon components.

- Low frequency modulation was found in the infrared emitted by qigong masters in the frequencies of 0.05, 0.15, 0.2 and 0.3 Hz.
- Infrared radiation from qigong master deviated from standard infrared radiation from black body by 5%-7% in the wavelength ranging from 3 to 5 microns.
- Magnetic fields in the range of 0.25 to 1.67 gauss were recorded for some qigong masters above acupoint PC8.
- Subsonic phonons with frequency in the range of 1-2 Hz and 9-10 Hz were found.
- Vibration of arteries of patients who received treatment increased more than 500%.

Q4.4.3. What does external qi mean in quantum field theory?

In the language of quantum field theory, sound is made up of phonons. Phonons exist in solids as well as in air. When phonons propagate in a solid we cannot hear them. When phonons propagate in air, we hear them as sound.

Qi are linear combinations of many quantum fields. These quantum fields may propagate along meridians. We call them internal qi. They may also propagate outside the human body, as external qi.

There are two kinds of quantum fields of qi, as we discussed in Chapter 2: bosonic qions and fermionic qions. Some qigong masters' external qi has a fragrant smell. Most likely this kind of external qi is made up of fermionic qions. The bosonic qions may consist of phonons and photons. Phonons can penetrate metals, whereas photons cannot.

The qigong master who taught me qigong could diagnose human diseases. My brother learned qigong and at the height of his power could do that as well. It is quite possible they obtained diagnostic information about the person who was sick by using bosonic qions to scan them. The bosonic qions of the master interact with the qions of the person being scanned, and give the master information about the sick person's internal organs. This is similar in principle to modern ultrasound scanning.

Q4.4.4. Can you propose some experiments to measure the properties of external qi?

If we assume that external qi comes from oscillations of the meridians, then there should also be external qi coming from acupoints of ordinary people, although the strength would be much weaker. The physical instruments for measuring photons and phonons need to be sophisticated enough to measure a strength of less than 1% of what has been measured for the qigong masters.

We propose to measure the orderliness of the emitted photons and phonons from acupoints of ordinary persons. Specifically we want to measure the infrared radiation emitting naturally from their skin. There are three properties of infrared radiation that would not be present in a random system.

1. Polarization of infrared radiation.
   For any electromagnetic waves emitted from an electric dipole, the radiation is polarized. A small amount of polarized infrared radiation emitting from one acupoint and not from a neighboring point would be definite evidence for the existence of electric dipoles in acupoints.
2. Deviation from random radiation.
   For any heat source at a particular temperature, there is a natural distribution of infrared radiation emitted, according to the Boltzman-Stefen law. Any deviation from such random radiation would indicate non-random radiation. Such non-randomness could only come from an ordered system like the meridian system we talk about here.

3. Coherent infrared radiation.
   If such a coherent component is found in the infrared radiation from acupoints of ordinary people, then it suggests the quantum effect is present. It then becomes necessary to use quantum field theory to thoroughly describe the phenomena of qi.

Q4.4.4. For ordinary mortals supreme health is the most one can aim for. But for ambitious people it is not enough. More ambitious people generally want to be in a spiritual state. Can we understand such spiritual states, if in fact they exist?

Our quantum scheme for human existence allows for many excited states, some of which could correspond to spiritual states. (See Chapter 2.)

Let us focus on one phenomenon: the aura surrounding the head. In the West, paintings of saints in the Middle Ages usually portrayed a glowing aura surrounding the heads of the saints. In the Orient, paintings of the Buddha often show an aura around his head.

In quantum theory, an electron would emit and absorb photons continuously. These photons form a cloud around the electron. The correct picture of an electron is a point-like object with a cloud of photons surrounding it. The diagram that illustrates this kind electron we call a self-energy diagram, shown in Fig. 4.4.4. The presence of such a cloud of photons has been confirmed by experiment.

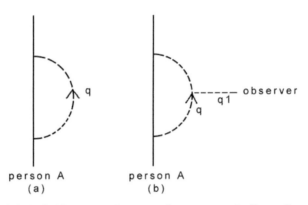

Fig. 4.4.4. Self energy diagram of a person, A. He emits qion q and receives it back, as shown in (a). When the qion is strong enough to emit another qion q1, then an observer may observe the qion cloud surrounding A through detecting q1.

It is quite clear if stable water clusters exist and line up along the meridians and acupoints, then a photon cloud would also be possible. They are likely to be the cloud-like halo, or auras, around persons such as the saints of the Middle Ages. In principle we should be able to detect these clouds using modern scientific equipment.

Chapter 5.

Standing on Solid Ground – Philosophy and Mathematical Formulations

Philosophy and quantum theory provide a solid foundation with an elevated perspective for the future of health

Taichi diagrams – a dual analysis of opposites

Q5.1.1. Traditionally it is said that a good Chinese doctor must be versed in philosophy as well as medicine. Why is this so?

In Western medicine, treatment is usually standard once the conditions of the patient are known. There is no room for philosophy. In Oriental medicine, each treatment is customized for the best result. No two patients are the same, so no two treatments are exactly the same. The person is treated at the most complex, and at the same time, basic level. Well-known methods derived from exact science are often insufficient to deal with great complexity. It is sometimes necessary to appeal to the basic laws that govern everything in the universe. These basic laws are a fundamental part of Oriental philosophy. Philosophy, in this sense, is useful to a good doctor of Oriental medicine.

Q5.1.2. The concept of taichi, or duality, originated about a thousand years ago. It is one of the more profound philosophical concepts ever formulated. How is it relevant to quantum theory, which is currently the basis of our meridian theory?

When quantum theory was developed, it contradicted many old concepts in physics. Niels Bohr proposed the concept of duality to reconcile many of these inherent contradictions. Duality means the complementary coexistence of two polar opposites. These opposites continually produce, imply, interpenetrate and interact with each other. This view of life coincides with the Asian concept of yin and yang. Duality, seen as yin and yang, can best be represented by a taichi diagram.

Before the invention of quantum theory, waves and particles were not seen as being the same. Now we understand that a wave is the motion of particles. For example, a wave in water is a movement of particles. The particles of a water wave are water molecules. Waves are the motion of particles. Particles are matter. The properties of waves and particles coexist together. Waves are matter, and particles have wave properties. Electrons are particles, but also have wave properties. Electromagnetic waves are waves, but are a collection of particle-like objects, photons

Q5.1.3. Can you give another example of the coexistence of the opposites in quantum theory?

The coexistence of certainty and uncertainty is another set of opposites in quantum theory. Conservation of energy and the momentum of any reaction are exact, and not uncertain. But the energy level of an atom or the momentum of an electron is uncertain. The amount of uncertainty is governed by Heisenberg's uncertainty principle.

Q5.1.4. Are there any fundamental laws in philosophy that are universally applicable to everything in the universe?

Yes, fundamental laws in philosophy do exist. Here are two:

1. Wuchi becomes taichi. Conversely, taichi returns back to wuchi (Fig. 5.1.3). Wuchi may be described as the undifferentiated, unstructured aspect of any system or subsystem. It has structure, but only at a more microscopic level of analysis or observation. Each level of existence may be seen as having subsystems, which are whole in regard to their parts, and parts with respect to the larger wholes. Any unit or phenomenon may be divided into two, and two may again be seen as one, depending on the level of observation or analysis. A hydrogen atom may be seen as a wuchi, but then it may be split into positive and negatively charged particles, becoming two taichi. Wuchi is represented by a circle.

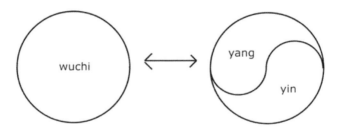

Fig. 5.1.3. Wuchi becomes tiachi, and taichi can return back to wuchi. Such is the fundamental property of wuchi and taichi.

2. One taichi splits into two wuchi. Conversely, two wuchi recombine to form one taichi. This is to say that one taichi, seen as separate from its opposite, may be seen as a whole in itself – a wuchi. So, for example, when a hydrogen atom is split, the separate electron and proton are each wholes in themselves (Fig. 5.1.4)

3. By repeatedly using these two fundamental laws we can reproduce all the taichi diagrams drawn in this book.

Numerous Chinese philosophers over the last thousand years have thoroughly and repeatedly discussed these two laws in different forms. We apply them here explicitly to explain the quantum theory of meridians.

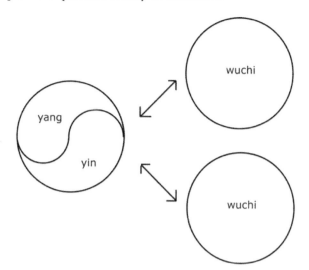

Fig. 5.1.4. One divides into two. A taichi splits into two wuchi. Two wuchi can rejoin back to become one taichi.

Q5.1.5. Taichi is putting yin and yang together. In simple diagrammatic form, it is putting white and black together. Is there any subtlety to it?

The subtlety is that there is yin in yang and there is yang in yin. In diagrammatic form there is white in black and there is black in white. There is never pure white or pure black (Fig. 5.1.5.). For simplicity and convenience we draw all taichi diagrams without such subtlety, as simplified versions of the proper taichi. It is important to remember these subtleties when we read and interpret the simplified taichi diagrams.

Q5.1.6. What have you done with these taichi diagrams that is new?

We have done basically two things. First, we have used the general concept of yin and yang to represent any two opposites. Second, we have repeatedly used these taichi diagrams. They then form a complex pattern that can illustrate the interrelations of a complicated object.

If we take away the subtlety in a taichi, there is no yin in yang and there is no yang in yin. Yin is yin, and yang is yang. Then we have a binary system: 0 can represent yin and 1 can represent yang. The binary system is the simplest number system, which is, however, useful in certain applications.

If we put the subtlety back in taichi, then we no longer have a binary numerical system, but a qualitative analytic system. Such a system can be used for a dual analysis. So we may call this way of using taichi diagrams a dual analysis of two opposites. It is a qualitative, not a quantitative analysis.

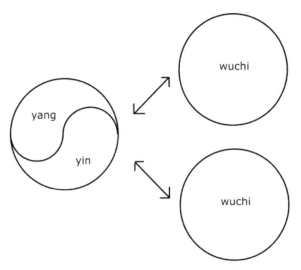

Fig. 5.1.4. One divides into two. A taichi splits into two wuchi. Two wuchi can rejoin back to become one taichi.

Quantum fields for bosons

Q5.2.1. Why do we need quantum fields?

A physical theory is invented to describe physical processes. In a classical point of view, an elementary particle is an indestructible object. It is neither created nor is it destroyed. The classical picture changes when we realize the light that comes from a light bulb is made up of elementary particles – photons. Since in a light bulb, light is being created continuously, then light disappears continuously when it shines on a piece of black cloth. Elementary particles are being created and destroyed continuously. A proper physical theory has to incorporate the process of creation and annihilation. The theory of quantum fields was invented to do this.

Q5.2.2. What does a quantum field consist of?

Since in quantum theory particle and wave properties coexist, a quantum field must contain particle properties and wave properties: a particle travels in space with momentum **k** and energy ω is created by a creation operator a^+_k. This creation operator operates in a vacuum to create a particle of momentum **k**:

$$a^+ \mid 0 > = \mid \mathbf{k} >,$$

where the vacuum is represented by the state vector $\mid 0 >$, and the particle is represented by the state vector $\mid k >$. The annihilation operator a_k annihilates a particle to vacuum with momentum k:

$$a_k \,|\, k \rangle = |\, 0 \rangle.$$

A wave that travels in space with volume V at momentum **k** and energy ω, on the other hand, is described by a plane wave function:

$$\exp(-i\,k\,x) = \exp(-i\mathbf{k}\,\mathbf{x} + \omega t)/\sqrt{V}.$$

The wave number and frequency are the same as the momentum; and energy, in the case when the Planck constant is set to unity. Then the quantum field Φ for bosons is a linear sum of all plane waves with all possible energy and momentum in space with volume V:

$$\Phi = \Sigma\,(a^{+}_{k}\,\exp(-i\mathbf{k}\,\mathbf{x} + \omega t)/\sqrt{V} + a_{k}\,\exp(i\mathbf{k}\,\mathbf{x} - \omega t)/\sqrt{V})$$

For a finite space with volume V, momentum and energy take discrete values, while space and time run continuously (F 5.2.3-4).

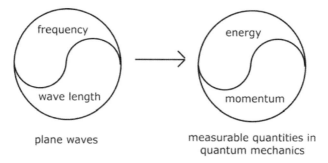

Fig. 5.2.3. Plane waves: The variables of plane waves are frequencies and wave lengths. With a multiplicative factor of the Planck constant, they become energy and momentum, which are measurable quantities.

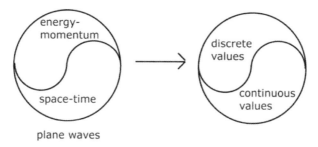

Fig. 5.2.4. Plane waves consist of four variables: energy, momentum, space and time. Energy and momentum have discrete values, and space time has continuous values in restricted space.

Electrons, fermionic qions and symmetry

Q5.3.1. What is the difference between electrons and photons in their mode of existence?

Photons, the elementary particles of light, can be absorbed by any light-absorbing materials and disappear without a trace. They can be created from out of a vacuum. Electrons, on the contrary, can be absorbed by a piece of metal, but they do not disappear. They just travel to another part of the metal. Electrons cannot be created from a vacuum. However they can be created with positrons, their anti-particle. Positrons have positive charges, whereas electrons have negative charges. Positrons can annihilate electrons. Then both positrons and electrons vanish together. Anti-matter, in general, annihilates matter completely. Electrons are fermions with half integer spins, and photons are bosons with integer spins. Photons are described by quantum fields, similar to those we discussed earlier in this chapter in Q5.2.2. Electrons are described by a different kind of quantum field.

Q5.3.2. The mathematical expression of quantum fields for fermions, like electrons, should contain all such differences. Does it?

Yes, the quantum field ψ for half spin fermions like electrons should contain the creation operator d^+ for the positron and the annihilation operator b for the electron. Its Hermitian conjugate ψ^+ should contain the annihilation operator d for the positron, and the creation operator b^+ for the electron.

The quantum field for electrons is

$$\Psi = \Sigma \ (d_p^+ \exp(-i\mathbf{p} \cdot \mathbf{x} + iEt) \ v^+(\mathbf{p}) + b_p \exp(i\mathbf{p} \cdot \mathbf{x} - I Et) \ u(\mathbf{p}))/ \sqrt{V}$$

where **p**, E, are momentum and energy of either the electron or the positron, V the normalization volume. The summation Σ is the over-all spin state and all allowed states of momentum and positive energy. The spinor wave functions $v^+(\mathbf{p})$ and $u(\mathbf{p})$ are for the positron and electron respectively.

Q 5.3.3. It is necessary to have explicit forms of the quantum field of electrons and photons to do calculations with the reactions among electrons and photons. However, when we deal with humans, we are dealing with an entity much more complex than elementary particles. The precise form of quantum fields may not be known for a long time. Are there general features, such as symmetries of quantum fields, which are relevant in complex objects?

Yes, there are three fundamental symmetries associated with quantum fields: symmetry in space, time, and between matter and anti-matter. Charge conjugation operator C turns particle into antiparticle, and changes the sign of charges. Time reversal operator T changes time t into –t. It means time runs backward. Parity operator P changes left to right and right to left. It is a reflection in space (Fig. 5.3.2-3).

The CPT* symmetries are particularly important, because any interaction one can write for quantum fields is invariant under the CPT transformation. So these symmetries may be very important in systems as complex as humans, as we discuss briefly in Chapter 3.

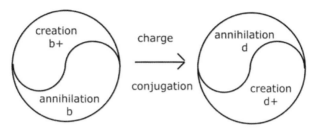

Fig. 5.3.2. Charge conjugation changes electrons into positrons.

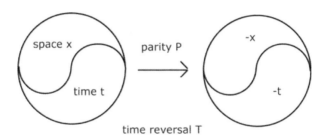

Fig. 5.3.3. Parity operation and time reversal operation changes x to -x and t into -t.

---

*CPT theorem: fundamental ingredient in quantum field theories, which dictates that all interactions in nature, all the force laws, are unchanged (invariant) on being subject to the combined operations of particle-antiparticle interchange (so-called charge conjugation, C), reflection of the coordinate system through the origin (parity, P), and reversal of time, T. If an interaction is not invariant under any one of the operations, its effect must be compensated by the other two, either singly or combined, in order to satisfy the requirements of the theorem. (McGraw-Hill, *Encyclopedia of Physics*, second edition, 1993)

Dramatic approximation for a human being

Q5.4.1. We are busy enough using classical physics and biochemistry to explain the reactions inside human body. Why do we need to introduce quantum field concepts to describe human beings? Isn't it making human beings more complex to understand?

We are forced to use the concept of quantum fields if we want to describe life. Life is about creation, and the only entity that incorporates creation is the quantum field. Quantum fields are adequate to describe the evolution of the biggest known object, the universe, as well as the intricate behavior of the smallest known objects: quarks and electrons. It is quite clear that proper use of quantum fields should be able to describe the crucial behaviors and cast great light on understanding human behaviors. On one hand, quantum fields are complex, as are human behaviors. But on the other hand, we have to make an enormous simplification of human beings in order to apply the concept of quantum fields.

Q5.4.2. What kind of simplification do we need? Is there more than one kind of simplification?

There are different kinds of simplification of a human being for different applications. As a result we have different quantum fields for each simplification:

1. Quantum fields for qions:

If we are interested only in qi, it is sometimes sufficient to quantize qi itself, and neglect all the rest. We may do it in this manner: From the study of external qi emitted by qigong masters, we know that the phonon is a component of a qion. Phonons are bosons. So a qion contains a component of bosons, which we call $\eta(x)$. A qion also contains a component of fermions, which is $\xi(x)$. A general form for the quantum field $q(x)$ of a qion is a linear combination of bosonic qions η_i and fermionic qions ξ_i.

$$q(x) = \Sigma \ (a_i \ \eta_i(x) \ + \ b_i \ \xi_i(x)), \qquad \text{(Fig. 5.4.1)}$$

where a_i, b_i are some constants and the sum Σ is over index i on all bosons and fermions.

In solid states, when we calculate the contribution of phonons to the specific heat of a crystal, we focus only on the phonons and neglect all other aspects of the crystal. Here it is the same. When we want to study one aspect of qi we may focus our attention only on qions and neglect all other aspects of a human being.

2. Quantum fields for the physical body:

a. **Using two layers of approximation:**
   In this model, we consider the physical body P of a human being as being made up of molecules only. Molecules are described by quantum field φ. If there are m molecules in the body P, then the quantum field Ψ for P is given by

   $$\Psi(x) = \Pi_i \varphi_i(x) \quad i \quad (5.4.2)$$

   where $i = 1, 2, 3 \ldots m$ and Π is the multiplication over all molecules. All molecules are assumed to pile on top of one another. They all have the same space-time distribution.

   A more realistic approximation is for the molecules to spread out, and the resultant quantum field is an integral of each quantum field weighted by a distribution function f_i:

   $$\Psi(x) = \int \Pi_i d x_i \ f_i(x, x_i) \varphi_i(x) \ldots \ldots \ I \quad (5.4.3)$$

   The distribution function f indicates the position of the molecules inside the physical body.

b. **Using four layers of approximation:**
   In this approximation we consider the human being as having four layers: the whole physical body, its molecules, cells, and molecules. There are l organs, n cells, and m organs inside a physical body.

   The physical body is made of l organs

   $$O_1, O_2, O_3, \ldots \ldots O_l \quad (5.4.4)$$

   The quantum field Ψ for the physical body P is given by

   $$\Psi = \int \Pi_i d x_i \ g_i(x, x_i) O_i(x_i), \quad (5.4.5)$$

   where $O_i(x)$ is the quantum field for organ O_i, and the multiplication Π is over all organs with $i = 1, 2, 3, \ldots l$. The distribution function g(x) indicates the distribution of the organs inside the physical body. The quantum field of the organ O_i is given by

   $$O_i(x) = \int \Pi_j d y_j \ h_j(y, x) C_{ij}(y_j), \ldots J \quad (5.4.6)$$

   where C_{ij} are quantum fields for cells inside the organ O_i, with $j = n_{i-1} + 1, n_{i-1} = 2, \ldots$ to n_i. The distribution function h indicates the positions of the cells inside the organ. The sum of the number of cells in each organ

$$\Sigma n_i = n \qquad (5.4.7)$$

is equal to the total number cells inside the physical body.

The quantum field of cell C_{ij} is given by

$$C_{ij}(y_i) = \int \Pi \, dz_k \, F_i(y_j, z_k) \, \varphi_k(z_k), \ldots \ldots \, K \qquad (5.4.8)$$

where φ_k are the quantum fields for molecules, and $k = m_{ij-1} + 1, m_{ij-1} + 2, \ldots m_{ij}$. The multiplication Π is over all molecules in cell C_i. The distribution function F indicates the position of molecules inside the cell C, the sum of all molecules

$$\Sigma m_{ij} = m \qquad (5.4.9)$$

Q5.4.3. What are the most crucial aspects of these formulae?

As in quantum theory the most crucial aspect is perhaps the quantum number of the system. For a two-layer approximation, the only crucial number is the number of molecules m:

$$P(m)$$

For a four-layer approximation, we have three total numbers: l, m, n. There are l organs, n cells, and m molecules in the body. The physical body P is characterized by three numbers:

$$P(l, m, n) \qquad (5.4.10)$$

The n should be subdivided into the number of cells n_i in each organ C_{ij}, and the number of molecules, m_i, should be subdivided into each cell m_{ij}.
Then we should consider the time derivatives of these numbers:

$$dl/dt = 0, \, >0, \, <0 \qquad (5.4.11)$$

The time derivative of the organ number l is positive when the person is in embryo stage and organs are formed. All organs are present at birth and the number l reaches a stable number, and $dl/dt = 0$. When the person grows old, his eyes may go blind, the number of functioning organs may decrease, and the derivative $dl/dt < 0$.

Similarly, in each organ O_i we are interested in the time derivative of the total number of cells n_i:

$$dn_i/dt = 0, \, <0, \, >0. \qquad (5.4.12)$$

After birth each organ is growing. The time derivative of each n_i is greater than zero. When the person reaches maturity, the number of cells become stable, and can be approximately set equal to zero. As the person gets old, the cells in each organ die and are not replaced, the time derivative of n_i becomes negative.

In each cell C_{ij} we are interested in the time derivative of the total number of molecules m_{ij}

$$dm_{ij}/dt = 0, <0, >0 \qquad (5.4.13)$$

When a person reaches maturity, some of his cells may get fat, then the time derivative of the number of molecules in the fat cell becomes positive. When a person's weight stabilizes, the number of molecules becomes constant. When he becomes sick and loses weight, the number of molecules decreases, and the time derivative becomes negative.

Q5.4.4. The electron as represented in the quantum field, in the above discussion of quantum fields for fermions and CPT symmetries, is for traveling plane waves only. What happens to electrons that are bound in atoms?

The electrons in atoms do not have discrete and conserved values in linear momentum, and cannot be represented by plane waves. They do have discrete and conserved values in angular momentum. So their quantum states are represented by the principal quantum number n, which denotes the size of the orbit. The smaller n is, the smaller the orbit. There are two types of angular momentum. The first is orbital angular momentum l, which indicates the shape of orbits electrons have. The second is total angular momentum, which is a sum of orbital angular momentum and the intrinsic spin of the electron.

In Fig. 5.4.1 we list some of the states for the hydrogen atom. It is interesting to note the method of approximation. In the far right of the figure, we use a gross approximation, and show the three different energy states with three different sizes of orbits: n=1, 2, 3. When we get to know the detail of the shapes of the orbits, we need to insert other quantum numbers to differentiate different kinds of orbits. Different values of total angular momentum j, and different orbital angular momentum l will give different states. We call this the splitting of states. We show in the figure the splitting of the state n=2 into three different states with three sets of different angular momentums. Each such state can be split further under the influence of a magnetic field. The state with the same angular momentum will split further according to different values of the projected angular momentum along the direction of the magnetic field. For a j=3/2 state, it will have four projections: J= 3/2, 1/2, -1/2 and -3/2. This is shown in the far right of Fig. 5.4.1. We split the states more as we know more and can measure more.

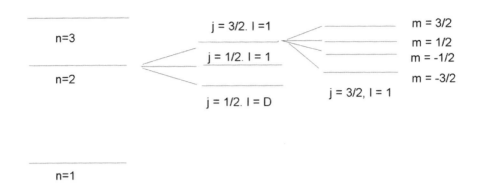

Fig. 5.4.1. Quantum states of hydrogen atom. The far left contains three energy eigenstates with different principal quantum numbers n. Each state is split into several states, depending on their total angular momentum j and orgital angular momentum I. We show the splitting of different angular momentums in the middle. Under the influence of a magnetic field, each state will split further into different states with a different quantum number m. We show the splitting of j = 3/2, I = 1 into four states of different m values.

Q5.4.5. This concept of splitting of quantum states looks like the right one to extend to quantum states of human beings. Is it true?

Yes, we could use the approximate methods to classify different states of the human being, depending what we want to know and how much we will be able to measure in the future.

The simplest three quantum states are the physical body P, the ground state I_0, and the excited state I. The excited state I is what we are in when we are awake and doing things. It has the highest energy. The physical body P is what the body consists of without life and functioning. It has the lowest energy state. In the middle is the ground state I_0, where a person is just alive, but not doing anything.

For the ground state there is a possibility that the bosonic qions would all become one single wave function, as usually happens in the ground state of superconductivity. Then the meridians would be described by one giant wave function that reaches to every part of the body. If there is an electric current flowing, it would be flowing without resistance. The body would be like a very sensitive detector that can receive very faint signals from everywhere, signals perhaps from far away. The onset of such a ground state can be determined by measuring the impedance among different acupoints of the meridians. There would be a sudden drop in impedances in all pairs of acupoints in any meridian. The signature is extremely sharp and unmistakable.

When we want to know more, we need to split these three levels to represent our knowledge. The ground state can be split into three states: First, the closest state to death, a person under total anesthesia, occupies the lowest split state. Second is the sleep state. Third, the most active of the ground states is probably the meditation state. It is well known from measuring brain waves, heartbeat, breathing, etc., that the meditation state is different from the sleep state. It is not clear whether the meditation state is of higher or lower energy than the sleep state.

The state of sleep has been well studied by measuring brain waves, eye movements, etc. It is not clear how many levels this state can be split into. Most people can distinguish at least two: deep sleep and shallow sleep. Deep sleep is sound sleep, after which one feels totally refreshed. Shallow sleep is bad sleep, after which one still feels tired. The more refined our measurements become, the more states of sleep will likely be found.

The excited state is when a person is conscious and active. It is not clear how many states this can be split into. Most people agree there should be at least two: the state where the mind is actively functioning, and the state of spirit. So in a crude approximation these two states are what people consider as the mind and the spirit of the body. As we are able to study, measure and understand brain waves better, we'll be able to split the conscious state into many states. These states, which are currently under great speculation, and are objects of study in philosophy, religion, and psychology, will be characterized by precise physical measurements.

The state of spirit is controversial. Everyone seems to have his own view of what it is. We might say the highest state a qigong master can attain is the state of spirit. In this state, he can emit external qi, which is measurable. The ability to measure the self-energy, as discussed in Chapter 4, will put some objectivity to the state of spirit.

Q5.4.5. What is the relation between quantum states and quantum fields?

Let us consider the electron. When an electron at ground state absorbs a photon, it jumps to an excited state. So the quantum field of photons turns the ground state of the electron into an excited state:

$$| \text{excited state} > = A | \text{ground state} >,$$

where $A(x)$ is the quantum field of photons. (5.4.14)

Similarly, in the simplest case, we may say that an excited state of a person is caused by the operation of the quantum field of $q(x)$ on his ground state:

$$| \text{excited state} > = q | \text{ground state} > \quad (5.4.15)$$

The quantum field of electron ψ that we introduced above can turn a vacuum state into an electron:

$$| \text{electron} > = \Psi | \text{vacuum} >$$

Replacing the quantum field of an electron with the quantum field ψ_p of a human, in our simple construct, can create the physical body from a vacuum. (For clarity we add a subscript P to designate the quantum field belongs to a person).

$$| P \rangle = \Psi_p | \text{vacuum} \rangle$$

To turn a dead body P into the ground state of a living person, there is an operator G.

$$| \text{ground state} \rangle = G | \text{vacuum} \rangle \qquad (5.4.17)$$

The operator G should at least contain some qions and some operators that create brain waves, heartbeats, breathing, etc.

Survival, family and the gauge principle

Q 5.5.1. Can you express the desire to survive in mathematical form?

One of the most fundamental desires of a person is to live longer and not to die. The desire to survive is universal for all living objects, as well as for non-living objects.

An electron survives. In our daily encounters, except where there are nuclear reactions, electrons do not disappear. The property of the survival of electrons is expressed as a conservation law. In its simple version, it just says that the number of electrons do not change in any interaction. In more sophisticated terms, we say that the law of conservation of electron numbers means that the sum of electron numbers of all particles in any reaction remains the same.

The conservation law of electron numbers is broken in some grand unification scheme of particle physics. It is replaced, sometimes, by a more general law of conservation of fermion numbers.

Similarly, the survival of human beings can be expressed as a conservation law for a human number. A human number is simply the number of human beings. In any human interaction, it is expected that the number of human beings is conserved. The conservation of the human number is broken in war, in murder, or simply in natural death.

Q5.5.2. How do these number conservation laws express themselves in quantum field theory?

The conservation of these number laws express themselves as the gauge principle in quantum field theory.

Q5.5.3. What is the gauge principle?

Gauges are scales. When we draw a picture of a house, we change the scales. One meter of the house may be represented by only 1 cm on paper. The scale is changed by a multiplication factor. The shape of house, however, should not be changed by the change of the scale. Then the shape of house is said to be invariant under a change of scale.

For a quantum field ψ of an electron with spin ½, a gauge transformation is also a scale transformation when we multiply the quantum field with a complex number with only a phase exp (iδ), where δ is any constant. For any interaction that preserves the number of electrons, where a creation operator must be followed by destruction operator, or vice versa, the quantum field Ψ must enter with its Hermitian conjugate Ψ^+. Then the gauge transformation

$$\Psi \rightarrow \exp(i\delta) \Psi$$

will leave the bilinear form the same:

$$\psi^+\psi \rightarrow \psi^+\psi \tag{5.5.1}$$

The property of the bilinear form is that after it destroys a particle (or antiparticle), it will create back a particle (antiparticle), or after it creates an antiparticle (particle), it will destroy an antiparticle. The total number of particles remains the same. So the bilinear form of quantum fields preserves the electron number. Any action that that describes the interaction of an electron as containig bilinear form similarly preserves the electron number (Fig. 5.4.1).

It is also clear that the above argument works for any fermions, and so is valid for fermionic qions.

Q5.5.4. How does the gauge principle apply to the survival instinct of a human body?

Let us take the simplest two-level approximation of a human body, where we assume the body is made up of only molecules. The quantum field ψ for a human body is a simple multiplication of quantum fields of all the molecules. Assuming a steady state, where the number of molecules remains the same, the action consists only of bilinear fields in ψ. Since any bilinear field does not change under a gauge transformation, the action for a human body is invariant under a gauge transformation. So the number of humans is conserved, just like the number of electrons is conserved. Such conservation of the human body is reflected in the desire of humans to live as long as possible. The survival instinct can be explained as the result of gauge invariance.

Q5.5.5. This seems reasonable. Is it also possible to apply the gauge principle to parts of a human body?

Yes. Let us take a four-level approximation of a human body, where the body is made up of organs, cells and molecules. The application of gauge invariance to their actions implies conservation of the number of organs, the number of cells, and the number of molecules.

Alternatively, if gauge invariance holds, we can say that each organ has a life of its own and wants to live as long as possible. Accidents that take away an organ, say an eye, will violate the conservation law of organs. Each cell, also, wants to have a life of its own, and wants to live as long as possible. Death of a cell inside an organ violates the conservation law of the number of cells.

Molecules in cells also want to survive as long as possible, and the number of molecules is conserved. Such conservation of molecules will be broken when biochemical reactions occur. (See Table 5.5.1 for a summary of these results.)

Q5.5.6. There are gauge particles associated with gauge invariance. What are the gauge particles associated with human beings?

The gauge particle for the gauge invariance of electrons is photons. The gauge particles for living objects are not well studied. We do not know what they are. For a living human body, it is possible that some of the qions that circulate along the meridians are the gauge particle for the conservation of a living human. When a human dies, the number of humans decreases by one, and the conservation of the number of humans no longer holds. The qions that circulate along meridians in a living being no longer exist. This is similar to the situation when an electron disappears: the source for generating photons vanishes. There are no more photons. Gauge particles vanish when gauge invariance no longer holds. The gauge particles for organs, or cells, may also be qions. We do not know at present.

Q5.5.7. When people get married, everyone wants the marriage to survive. Can the survival of a marriage be related to the gauge principle?

The survival of marriage preserves the number of families. So the conservation number is the family number. To such a family number, we can associate a gauge particle and a gauge invariance. In daily life a couple functions very differently from two single people. It is natural to assign a doublet to a married couple. This would be similar to the doublet of an electron and its neutrino. Associated with such a double entity there will be a unitary group symmetry. The gauge group is SU(2). The fundamental doublets of this SU(2) gauge group then involve two human beings: a male and a female, or a husband and a wife. The couple forms a doublet.

It is quite clear that there exists great attraction between two persons who are in love with each other. Otherwise they would not be married. Attraction may be attributed to some kind of attractive force, like the attractive force between a positive charge and a negative charge. It is possible such an attractive force between two persons who have sexual intercourse may be biological in origin. Furthermore the biological origin may come from the physical existence of a gauge particle.

The gauge principle associated with SU(2) is well studied. It is called the non-abelian gauge, or the Yang-Mill gauge. Associated with doublets, then, there should be some gauge particles. The Yang-Mills gauge particle, theoretically, could apply to a doublet, or couple, in a family. It is not clear where to look for such a gauge particle, however. It is most likely that it could be some kind of qions as well.

In the qigong practice of some Taoist schools, it is important for a person to remain a virgin throughout his or her life. Any sexual intercourse will change the person's ability in a significant way. So the existence of a Yang-Mills gauge particle might be looked for as a difference in the qi that circulates inside the body of a virgin male or a virgin female before and after they have their first sexual intercourse.

The action principle, interaction and Feynman diagrams

Q5.6.1. We have been introduced to quantum fields and the gauge principle that governs their combination. The next question is how do they behave? How do they interact with one another?

If we have a bucket of water, and we pour the water on a floor, the water will flow to the lowest area of the floor. The lowest point is where water has its minimum potential energy. When we release a balloon, it will rise as high as it can go. Its maximum height is where its buoyancy force balances its weight. When we teach a young person, we always encourage him or her to achieve to the maximum potential.

The idea of an object, whether alive or not alive, wants to reach either maximum or minimum of "something" is elevated to a principle. This principle is called the action principle. This "something" is called the action S of the problem. The action S varies from problem to problem, and it is varied to achieve the extremal, which means either the maximum or the minimum of a given problem.

In classical physics, the action S for a particle is a function of kinetic energy and potential energy of that particle. By varying action S to achieve an extremal, an equation of motion is obtained. From the equation of motion we can calculate the future motion of any particle. The action S, as the name implies, is where all the action comes from.

A similar line of reasoning is followed in quantum field theory. In quantum field theory, the action S is a functional of energy density, which is called the Lagrangian. The Lagrangain is constructed from the quantum fields of a given problem. By varying action S to achieve an extremal, (either maximum or minimum) an equation of motion is obtained. From the equation of motion we can calculate the future of any interactions. Mathematically the action principle is formulated as a statement that the variation of the action δS is set to zero:

$$\delta S = 0 \qquad (5.6.1)$$

When we apply the action principle to living objects in ordinary daily situations, we have to consider all the constraints and boundary conditions that have been imposed externally upon the problem, we may say, to optimize the action to achieve the most optimal solution. It may not be the best we'd like to have, but given the constraints, it is the most optimal.

Q5.6.2. Many theoretical physicists regard the action principle as the most profound principle in physics. Can you elaborate on the structure of action?

The action principle is used as the starting point to derive formulae in classical mechanics, statistical mechanics, quantum theory, and general relativity. So it does have the qualification to be the most fundamental principle in nature. We use it to apply to human beings is obviously the next logical step.

The action S contains a yin and yang. The yang is the internal component. In quantum field theory it is the energy density called the Lagrangian. The yin is the external boundary conditions. Yin and yang occupy equivalent positions in a taichi. So explicit knowledge of boundary conditions is as important as the knowledge of internal components to obtain the correct solution.

There are many demonstrations of the effect of external qi by qigong masters. The qigong masters always claim that the effect will vary because the circumstances are different in each occasion. There may be some truth in it.

For experiments in physics one can always arrange the experiments so that the external constraints are similar in different experiments. Then one can utilize the same equation that is derived from the internal components for different experiments. In clinical tests in medicine the clinic is arranged in such a way so that it would be the same to different persons at different times.

Q5.6.3. For an object to be free on its own is quite different from when it interacts with others. How does action incorporate this feature?

The earth would travel in a straight line if it were free. The interesting situation is when earth interacts with the sun via gravitational force. Then the earth circulates around the sun to receive the heat from the sun, which is necessary for the existence of living objects.

Qi circulates freely around the meridians when the person is healthy. Nothing needs to be done. When a person becomes sick, we need to interact with the qi inside the body either by acupuncture or by medicine. These interactions are most interesting.

The action S is normally split into two sums: a term for the free object S_o, and a term for interaction S_I.

$$S = S_o + S_I \quad (5.6.2)$$

Let us now focus on the interaction term. For the interaction of an electron with a photon, the interaction Lagrangian is multiplying the bilinear fields $\psi^+ \psi$ of the electron field ψ with the photon field A:

$$e \psi^+ \psi A, \qquad (5.6.3)$$

where e is the charge of the electron. For simplicity reasons we have suppressed the spin structure of the electron and the photon. We want to emphasize that the electromagnetic interaction between an electron and any electromagnetic fields is completely defined by the above formula.

From this formula we can calculate all known electromagnetic processes, and the result agrees with experiments to a fantastic accuracy.

We are on a firm footing, as firm as it can get in the present knowledge, when we model the interaction of qi after the most successful electromagnetic theory as given by 5.6.2 above.

So for qions to interact with one another we take the simplest Lagrangian, which is also a multiplication of fields. There are the three qions interacting:

$$q^+ q \, \eta \qquad (5.6.4)$$

where q is the quantum field for qion. It is not possible for three fermion fields to interact together. The third field must be a boson field η of the qion. There are four qions interacting:

$$q^+ q q^+ q \qquad (5.6.5)$$

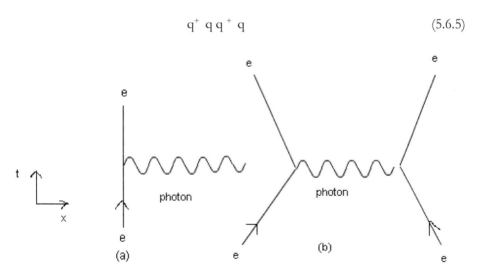

Fig. 5.6.4. Feynman diagrams for electromagnetic interactions: (a) An electron emits or abosrbs a photon. (b) The scattering of electron with another electron with an exchange of one photon. In the diagram above, the time dimension runs vertically; the space dimension runs horizontally.

If we expand the qion field q into its fermion field ξ and its boson field η, then from the two equations (5.6.4-5) we have the following interactions:

Three bosonic qion fields interact as the multiplication of three boson fields:

$$\eta^+ \eta \eta \qquad (5.6.6)$$

Four bosonic qion fields interact as the multiplication of four boson fields:

$$\eta^+ \eta \eta^+ \eta \qquad (5.6.7)$$

Two fermionic qion fields ξ interact with one bosonic qion field η:

$$\xi^+ \xi \eta \qquad (5.6.8)$$

Four fermionic qion fields interact with one another:

$$\xi^+ \xi \xi^+ \xi \qquad (5.6.9)$$

Two fermionic qion fields interact with two bosonic qions:

$$\xi^+ \xi \eta^+ \eta \qquad (5.6.10)$$

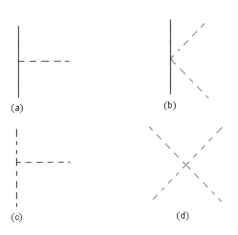

Fig. 5.6.5. Feynman diagrams for (a) vertex of f-qion with one b-qion, (b) vertex of f-qion with two b-qions, (c) vertex for three b-qions, (d) vertex for four b-qions. We use solid line to represent f-qion and dotted line to represent b-qion.

Q 5.6.4. Feynman diagrams of a particular reaction are a convenient way to represent the underlying physical process. We have used them in our previous chapters. How do those diagrams come from the interactions discussed above?

Let us describe an electron e traveling in one-dimensional space. Its trajectory will trace a line in one-dimensional space and one-dimensional time). At some space-time it emits a photon, and then it changes direction. This is the space-time diagram of interaction shown in (5.6.3). A fermionic qion emitting a bosonic qion, as described by the interaction (5.6.8) would be shown in the same way.

When we describe a more complicated process, like two electrons scattering off each other by an exchange of photon, we can trace the space-time trajectory of the two electrons in a one-dimensional space and one dimensional time diagram. The diagrams in Figs. 5.6.1, 5.6.4 and 5.6.5 are called Feynman diagrams. These are used extensively in quantum field theory to represent visually what happens among these very tiny particles.

Dipole radiation; exponential law in pain relief

Q5.7.1. If meridians are made up of a similar kind of molecule as surrounding tissue, the only difference is that molecules in meridians are more orderly. How can we prove it?

One way to prove that molecules making up meridians are more orderly than surrounding ones is to look for some order from the radiation they emit. Normal infrared radiation from an inanimate object is completely random and without order. The radiation from aligned dipoles is measured.

Q5.7.2. How do you measure non-randomness in infrared radiation from a warm object like a human body?

The infrared radiation from a black body (a body which emits the maximum amount of heat radiation), is random. The infrared radiation from a human body is known to be close to that from a black body but not exactly the same.

It has been shown that certain parts of the body of a qigong master can emit infrared radiation significantly different from black body radiation during his practice of qigong. The difference could be as much as 7%. The differential energy spectrum dE as a function of differential of angular frequency ω for black body radiation is:

$$dE = \text{const } \omega^3 \exp(-h\omega/2\pi kT) \, d\omega$$

where h, k are Planck's constant and Boltzman's constant respectively, and T is temperature in absolute scale K^0. The infrared radiation intensity I(D) over a bandwidth D, which is between wave lengths λ_1 and λ_2, is the integration of the energy over this bandwidth:

$$I = \int_D dE$$

$$D = (\lambda_1, \lambda_2).$$

So I depends on bandwidth and temperature. The integration can be performed exactly, and then the inverse relation can be easily obtained: the temperature distribution becomes a function of intensity I and bandwidth D:

$$T = T(I,D).$$

If a human body has a perfect black body radiation without any radiation that comes from some preferred processes, then the temperature T is independent of the bandwidth one uses. One should get the same temperature at the same spot from measuring different frequencies with different bandwidths of the infrared radiation.

It is quite clear that a human body is not a random object. It is living, and it is ordered. The infrared radiation from the surface must be different from black body radiation. And some part of the surface has a different distribution of infrared radiation than other parts. The Oriental meridian theory indicates that acupoints would emit more ordered radiation than other parts of the body.

Let us give an explicit numerical example. The normal infrared camera takes on the radiation bandwidth D from 8-14 microns, which will yield a surface temperature distribution T(x,y,z). A filter, which has a bandwidth D_1 from 3-5 microns, is used and intensity I_1 is measured. The temperature obtained is T_1. The spatial distribution of the difference of the two temperatures:

$$\Delta T(x,y,z) = T(x,y,z) - T_1(x,y,z)$$

is calculated and displayed in the monitor. The display of this difference could be in color code for different numerical values. Or, simply, if the difference in temperature is smaller than a designed value, say 0.05° C, which can be varied to emphasize different degrees of deviation from black body radiation, it is assumed to be zero. If it is above this designated value 0.05°, it is assumed to be 1, and is shown as gray spots on the ordinary thermal graphs, which display the spatial distribution of T(x,y,z) obtained from intensity I(x,y,z).

It is quite clear that the bandwidth of T_1 can be any other value with wavelengths between 8 and 10 microns as well.

This kind of measurement of ΔT can be done while the patient is undergoing treatment, or someone is practicing some special health exercise. It will show the effectiveness of the treatment or exercise. So $\Delta T(t,x,y,z)$ has both a spatial distribution x,y,z as well as a time distribution t.

Hence the detection of non-black body radiation can be used as a diagnostic and treatment tool as well as a health-monitoring device, such as keeping track of the effectiveness of a particular health exercise on the human body. Such health exercises can be qigong, aerobic, or training exercise for any sport. This invention can be applied in particular to sports medicine and competitive sports. It measures the particular effectiveness of a particular exercise to a particular individual at a particular time.

Q5.7.3. How do you detect the polarization of infrared radiation from acupoints?

An infrared polarizer is placed in front of the lens of an infrared camera. The polarizer may be made of crystal that transmits preferentially infrared radiation with a particular polarization state, and less so for its orthogonal state, or it may be made of wired grid which only allows polarized light, with polarization along the gap between two conducting wires, to pass through. The polarizer is caged in a holder that can rotate freely.

The polarizer is first set at a fixed direction, say in a vertical position, to allow polarization that is perpendicular to the horizontal plane to pass through. An infrared picture is taken at this position with an intensity of spatial distribution $I_1(x,y,z)$, where x,y,z are the spatial coordinates on the human surface where infrared radiation emits. Then the polarizer is rotated 90 degrees. An infrared picture is taken, and another intensity of spatial distribution $I_2(x,y,z)$ of the human surface is recorded. A software program in the computer connected to the infrared camera will process these two intensity distributions and calculate its normalized difference E_A, which is called the normalized polarization distribution function, as follows:

$$E_A(x,y,z) = (I_1 - I_2)/(I_1 + I_2)$$

If the infrared radiation is completely random, this normalized polarization distribution is zero everywhere. If there is any spot where the infrared radiation is not random, but is emitted with a definite polarization state (other than 45° off vertical) at the acupoint where it has been postulated to be populated by water clusters with permanent electric dipoles, then the radiation is polarized. The polarization function E will be nonzero at these acupoints. Display of the spatial distribution of the polarization E will show vividly the positions of the acupoints on the human surface. This distribution is displayed on a monitor, which can be shown together with the thermal distribution or with the visual images of the human body. The viewer can see where on the human surface the deviation from black body radiation is. A medical practitioner or an expert software programmer may understand the meaning of such distribution and make the diagnosis.

The spatial distribution of the polarization E can also be taken while the patient is under treatment. Its variation in time t also will provide additional information on the effectiveness of the treatment. So the normalized polarization function $E(t,x,y,z)$ will have an additional time dependence t. The function E may be shown as a contour or in color code to indicate its numerical values.

In many instances a simpler scheme may be used. The polarization function could be assigned to two values only:

$$E_A = 0 \text{ or } 1$$

It is zero if its absolute value falls below a designated value, say it is less than 1%, and it is assigned a value of 1 if its absolute value is bigger than this designated value, say 1%. When it is 1, it is represented by a black color. When it is 0, nothing is shown. So E_A may become a set of black spots. Black spots are places where polarized infrared radiation comes from.

The above treatment fails to address the case when the polarization if 45% off the vertical direction. More measurement is needed. The polarizer in front of the lens should then be rotated another 45% to take a picture with spatial intensity distribution $I_3(x,y,z)$. After the picture is taken, the polarizer should be rotated further to 90° to take another picture with spatial intensity distribution $I_4(x,y,z)$. A different normalized polarization distribution function E_B is defined as follows:

$$E_B(x,y,z) = (I_3 - I_4)/I_3 + I_4)$$

A new polarization function E can be defined as a sum of the above two polarization functions:

$$E(x,y,z) = E_A + E_B$$

As we discuss above, this function may be displayed in full numerical contour or color code with or without thermal distribution and visual images of the human body. There is also a simpler way to display that. This polarization E will be assigned a value 0, or 1, depending on whether its value exceeds a designated value, say 1%, which can be varied to emphasize the amount of polarized infrared radiation. If it is represented in black color on a thermograph derived from I, then the black spots show where there is polarized infrared radiation.

Q5.7.4. How can you derive an exponential law that seems to govern the decrease of temperature by acupuncture?

We do not know enough about the mechanism of pain in the human body to derive such an exponential law of decrease in temperature by acupuncture. Let us work backward, and take the exponential law as a given experimental fact and see if it can be reconciled with the hypothesis of meridians and acupuncture that is discussed in this book.

An exponential law in the decrease of temperature can be simply obtained if the decrease in temperature dT is proportional to its temperature:

$$dT = \text{constant} \times T \times dt,$$

where dt is the time changed. The solution to this differential equation is an exponential decrease in temperature:

$$T = T_0 \operatorname{Exp}(-\text{constant} \times t)$$

The temperature of the skin is related to the number n of biochemical reactions that occur per unit area. The relation in general can be expressed in the dependence of temperature as a power series of n:

$$T = a_0 + a_1 n + a_2 n^2 + \ldots$$

where a_0, a_1, a_2 are some constants. When there is no biochemical reaction n = 0, there is no heat generated, the temperature T is close to zero. So we may set a_0 to be zero. Neglecting all higher order terms, we have a linear relation between the temperature and the relevant number of biochemical reactions n:

$$T = a_1 n$$

The differential equation for temperature becomes the differential equation for n:

$$dn = \text{constant} \times n \times dt$$

When an acupuncture needle is inserted into an acupoint, additional qi is generated and propagated to the pain region. The interaction of qi with cells produces a curing effect on the cell so that biochemical reactions are reduced. As long as such interactions are described by quantum theory, it is statistical. The transition rate of each cell that is restored to normality is proportional to the number of cells present. The above equation is satisfied.

In Conclusion

Quantum field theory is needed to explain birth and death. It is the only physical theory that includes creation and annihilation operators. Further, one of its most fundamental principles, the gauge principle, describes the fundamental natural survival instinct of a human being.

Using quantum theory, the interaction of meridians and their organs, which traditionally were associated with five elements – metal, wood, water, fire and earth – have been formulated as the interaction of various qions. In classical physics, two colliding electromagnetic waves pass through each other without interacting. Only in quantum field theory can two electromagnetic waves, as photons, interact. So to describe the interaction of two bosonic qions from different meridians, the quantum field seems to be most appropriate. Furthermore, defining qi as a collection of qions in terms of the quantum field makes it measurable and subject to future experiments.

The purpose of life has been formulated as the action principle for the interaction of qions. Since meridians connect all organs, tissues and cells, a person's action derived from the action principle automatically involves all parts of the body in a coherent and holistic way, for the good of the complete entity.

Watercircles   by alpha lo

Water
  Neither solid, nor lacking structure,
  its form changing.

Circles
  No start, no finish,
  oscillating, resonating.

Light
  No past, no future.
  Bits created and annihilated.

Quantum
  Not here, not there,
  its vacuum pregnant with energy

From the alive nothingness
quantum oscillations arise,
creating lines of water cluster strings
with backbones of electric fields.

Meridians
The basis of bodily life
Alive with the hum of vibrations,
the dance of light,
the flow of information and ionic electricity,
conducting the quantum resonance
between acupoint and organ,
between cells and brain,
transporting energy from one part of the body
to another.

Molecules change vibration levels, emit light,
that is absorbed by another molecule,
influencing its behavior
in the symphony of the body.

The meridian is a circle,
light, current, information,
traveling around and around.
No start, no finish

References

Chapter 1.

A Universe within a Universe

Canon of Medicine, Chapter 74. First published in approximately 500 B.C.

Meridians

Stable Water Clusters

Shui-yin Lo and Wenchong Li. Onsager's Formula, Conductivity, and Possible New Phase Transition. Modern Physics Lett B, v.13(1999) 885-893.

Shui-yin Lo, Anomalous State of Ice, Modern Physics Lett B, v.10(1996) 909-919.

Shui-yin Lo, A. Lo, W.C. Li, Li T.H., Li H. H., and Xu Geng. Physical properties of water with I E structures. Modern Physics Lett. B, v. 10(1996) 921-930.

Shui-yin Lo and B Bonavida. Proceedings of First Int. Symp. of Physical, Chemical, and Biological Properties of Stable Water (I E) Clusters (World Scientific 1998) – (book)

Shui-yin Lo and W. C. Li and S. H. Huang, Water clusters in Life, Medical Hypotheses (2000)v 54(6), 948-953.

Shui-yin Lo and W. C. Li, Nanostructures in very dilute aqueous solutions, Russian Mendelev Journal of Chemistry 541.6:54-145.3, p41-48.

Shui-yin Lo, Hypothesis of meridians and infrared imaging. Accepted for publication in Medical Hypothesis.

A. Electrical properties

Tian JB, Shen S, Han JS. Frequency dependence of somatostatin and calcitonin gene related peptide release induced by electroacupuncture in rat spinal cord. Sheng Li Xue Bao. 1998 Feb;50(1):101-5. Chinese. PMID: 11324508. [PubMed - indexed for MEDLINE]

Hsieh CL, Kuo CC, Chen YS, Li TC, Hsieh CT, Lao CJ, Lee CJ, Li JG. Analgesic effect of electric stimulation of peripheral nerves with different electric frequencies using the formalin test. Am J Chin Med. 2000; 28(2):291-9.

Kwon Y., Kang M. Ahn C, Han H, Ahn B, Lee J. Effect of high or low frequency electroacupuncture on the cellular activity of catecholaminergic neurons in the brain stem. Acupunct Electrother Res. 2000; 25(1):27-36.

Hsieh CL, Lin JG, Li TC, Chang OY. Changes of pulse rate and skin temperature evoked by electroacupuncture stimulation with different frequency on both Zusanli acupoints in humans. Am J Chin Med. 1999; 27(1):11-8.

Guo HF, Wang XM, Tian JH, Huo YP, Hans JS. 2 Hz and 100 Hz electroacupuncture accelerate the expression of genes encoding three opioid peptides in the rat brain. Sheng Li Xue Bao. 1997 Apr; 49(2):121-7. Chinese.

Lin TB, Fu TC, Chen CF, Lin YJ, Chien CT. Low and high frequency electroacupuncture at Hoku elicits a distinct mechanism to activate sympathetic nervous system in anesthetized rats. Neurosci Lett. 1998 May 15; 247(2-3):155-8.

Guo HF. Comparative study on the expression and interaction of oncogene c-fos/c-jun and three opioid genes induced by low and high frequency electroacupuncture. Sheng Li Ke Xue Jin Zhan. 1996 Apr; 27(2):135-8. Chinese.

Guo HF, Tian J, Wang X, Fang Y, Hou Y, Han J. Brain substrates activated by electroacupuncture of different frequencies (I): Comparative study on the expression of oncogene c-fos and genes coding for three opioid peptides. Brain Res Mol Brain Res. 1996 Dec 31; 43(1-2):157-66.

Shen S, Bian JT, Tian JB, Han JS. Frequency dependence of substance P release by electroacupuncture in rat spinal cord. Sheng Li Xue Bao. 1996 Feb; 48(1):89-93. Chinese.

Zhang W, Xu R, Zhu Z. The influence of acupuncture on the impedance measured by four electrodes on meridians. Acupunct Electrother Res 1999; 24(3-4):181-8.

Huang X, Xu J, Wu B, Hu X. Observation on the distribution of LSIPs along three yang meridians as well as ren and du meridians. Zhen Ci Yan Jiu 1993; 18(2):98-103.

Yu C, Zhang K, Lu G, Xu J, Xie H, Lui Z, Wang Y, Zhu J. Characteristics of acupuncture meridians and acupoints in animals. Rev Sci Tech 1994 Sep;13(3):927-33

B. Photonic

Litscher G, Wang L.Thermographic visualization of changes in peripheral perfusion during acupuncture. Biomed Tech (Berl). 1999 May; 44(5):129-34. German.

Zhang D, Fu W, Wang S, Wei Z, Wang F. Displaying of infrared thermogram of temperature character on meridians. Zhen Ci Yan Jiu. 1996; 21(3):63-7. Chinese.

Zhang D, Gao H, Wei Z, Wen B. Preliminary observation of imaging of facial temperature along meridians. Zhen Ci Yan Jiu. 1992;17(1):71-4. Chinese.

Zhang D, Gao H, Wen B, Wei Z. Research on the acupuncture principles and meridian phenomena by means of infrared thermography. Zhen Ci Yan Jiu. 1990;15(4):319-23.

Liu R, Zhuang D, Yang X, Li Y, Zhang D, Wen B, Zhang R. Objective observation on phenomena of sensation along channels (PSC) and QI reaching to affects area (QIRA) – the influence of acupuncture points on infrared thermal image of face. Zhen Ci Yan Jiu. 1990;15(3):245-9. Chinese.

Liu R, Zhuang D, Yang X, Li Y, Zhang D, Wen B, Zhang R. Objective display on phenomena of propagated sensation along channels (PSC) – changes on the infrared thermal image channels pathway of upper extremity. Zhen Ci Yan Jiu. 1990;15(3):239-44. Chinese.

Osipov VV, Glibitskii MM, Kasatkin IL. Temperature oscillation spectrum at biologically active points and changes in them during activation of acupuncture meridians. Biofizika 1996 May-June; 41(3):749-54.

Choi C, Lee S.M., Yoon G., and Soh K. Propagation of light along acupuncture meridian, preprint.

C. Mechanical

Yang B, Li W, Hu X, Wu B, Li B, Li L, Chen J, Chen L, Zhang D, Xu W. Observation on the phenomenon of propagated sensation along meridians in youngsters. Zhen Ci Yan Jiu. 1993; 18(2):159-62. Chinese.

Li B, Li Lng L, Chen J, Chen L, Xu W, Gao R, Yang B, Li W, Li W, Wu B, et al. Observation on the relation between propagated sensation along meridians and the therapeutic effect of acupuncture on myopia of youngsters. Zhen Ci Yan Jiu. 1993; 18(2):154-8. Chinese.

You Z, Hu X, Wu B, Zhang W, Liang D. The difference in the time for the appearance of acupuncture effects between the subjects with and without PSM during acupuncture of neiguan. Zhen Ci Yan Jiu. 1993; 18(2):149-53, 148. Chinese.

You Z, Hu X, Wu B, Zhang W, Liang D. Experimental observation on the relation between pericardium meridian and cardiac function. Zhen Ci Yan Jiu. 1993;18(2): 143-8. Chinese.

Xu J, Huang X, Wu B, Hu X. Influence of mechanical pressure applied on the stomach meridian upon the effectiveness of acupuncture of zusanli. Zhen Ci Yan Jiu. 1993; 18(2):137-42. Chinese.

Wu B, Hu X, Yang B, Xu J, Li W, Li B, Chen J, and Chen L. The influence of pressing the meridian course on electroretinogram during acupuncture. Zhen Ci Yan Jiu. 1993; 18(2):132-6. Chinese.

Wu B, Hu X, Xu J, Yang B, Li W, Li B. Localization of the meridian track over body surface by the method of blocking the acupuncture effect with mechanical pressure. Zhen Ci Yan Jiu. 1993; 18(2): 128-31, 114. Chinese.

Wu B, Xu J, Hu X, Yang B, Li B. Observation on the functional characteristics of cortical somatosensory area during the advance of the propagated sensation along meridians. Zhen Ci Yan Jiu. 1993; 18(2): 123-7. Chinese.

Hu X, Wu B, Huang X, Xu J, Yang B, Gong S, Li B. Evidences for the appearance of peripheral activator during the advance of PSM, Zhen Ci Yan Jiu. 1993; 18(2):115-22. Chinese.

You Z. Preliminary observation on the relation among needling sensation, propagated sensation along meridian (PSM), and acupuncture effect when acupuncture neiguan. Zhen Ci Yan Jiu. 1992; 17(1):75-8. Chinese.

D. Ionic

Fei Lun, Cheng Huansheng, Cai Deheng, Yang Shixun, Xu Jianrong, Chen Eryu, Dang Ruuishan, Ding Guanghong, Shen Xueyong, Tang Yi and Yao Wei. Chinese. Experimental exploration and research prospect of physical bases and functional characteristics of meridians. Science Bulletin 1998 43(15):1233-1251.

Wang W, Zhang Y, Guo Y, Miao W, Wang X, Xutang P. The detection and studies on the change of H+ concentration in the regular points of the rabbit suffering from arrhythmia induced by aconitine. Zhen Ci Yan Jiu 1996; 21(4):59-63.

Acupoints

Properties of Acupoints

Chiou SY, Chao CK, Yang YW. Topography of low skin resistance points (LSRP) in rats. Am J Chin Med 1998; 26(1):19-27.

Cho SH, Chun SI. The basal electrical skin resistance of acupuncture points in normal subjects. Yonsei Med J 1994 Dec; 35(4):464-74.

Jakoubek B, Rohlicek V. Changes of electrodermal properties in the "acupuncture points" on men and rats. Physiol Bohemoslov 1982; 31(2):143-9.

Comunetti A, Laage S, Schiessl N, Kistler A. Characterisation of human skin conductance at acupuncture points. Experientia 1995 Apr 15;51(4):328-31.

Sher L. Effects of electrostatic potentials generated on the surface of the skin by wearing synthetic and semisynthetic fabrics on physical condition, mood and behavior: role of acupuncture points. Med Hypotheses 2000 Mar; 54(3):511-2.

Mustafin AM.The connection between the biologically active points of the skin and psychological functions. Biull Eksp Biol Med 1993 Jul; 116(7):100-1.

Poon CS, Choy TT, Koide FT. A reliable method for locating electropermeable points on the skin surface. Am J Chin Med 1980 Autumn; 8(3):283-9.

Shang C. Electrophysiology of growth control and acupuncture. Life Sci 2001 Feb 9; 68(12):1333-42.

Rakovic D. Neural networks, brainwaves, and ionic structures: acupuncture vs. altered states of consciousness. Acupunct Electrother Res 1991;16(3-4):89-99

Shang C. Singular point, organizing center and acupuncture point. Am J Chin Med 1989; 17(3-4):119-27.

Function of Acupoints

Radmayr C, Schlager A, Studen M, Bartsch G. Prospective randomized trial using laser versus desmopressin in the treatment of nocturnal enuresis. Eur Urol 2001 Aug; *40(2):201-5.*

Cramp AF, Noble JG, Lowe AS, Walsh DM. Transcutaneous electrical nerve stimulation (TENS): the effect of electrode placement upon cutaneous blood flow and skin temperature. Acupunct Electrother Res 2001; 26(1-2):25-37

Shi R, Ji G, Zhao L, Wang S, Dongjun. Effects of electroacupuncture and twirling reinforcing-reducing manipulations on volume of microcirculatory blood flow in cerebral pia mater. J Tradit Chin Med 1998 Sep; 18(3):220-4.

Branco K, Naeser MA. Carpal tunnel syndrome: clinical outcome after low-level laser acupuncture, microamps transcutaneous electrical nerve stimulation, and other alternative therapies – an open protocol study. J Altern Complement Med 1999 Feb; 5(1):5-26.

Schlager A, Offer T, Baldissera I. Laser stimulation of acupuncture point P6 reduces postoperative vomiting in children undergoing strabismus surgery. Br J Anaesth 1998 Oct; 81(4):529-32.

Chupryna HM. Infrared laser puncture in the treatment of facial neuritis. Lik Sprava 1997 Sep-Oct; (5):172-5.

Dong L, Yuan D, Fan L, Su L, Fu Z. Effect of HE-NE laser acupuncture on the spleen in rats. Zhen Ci Yan Jiu 1996; 21(4):64-7.

Sing T, Yang MM. Electroacupuncture and laser stimulation treatment: evaluated by somatosensory evoked potential in conscious rabbits. Am J Chin Med 1997;25(3-4): 263-71.

Macheret IeL, D'iachenko OIe, Korkushko OO. The treatment of patients with chronic cerebral circulatory failure by using laser puncture and the microclimate of the biotron. Lik Sprava 1996 Jul-Sep;(7-9):142-5.

Kanakura Y, Kometani K, Nagata T, Niwa K, Kamatsuki H, Shinzato Y, Tokunaga Y. Moxibustion treatment of breech presentation. Am J Chin Med 2001; 29(1):37-45.

Jacobsson F, Himmelmann A, Bergbrant A, Svensson A, Mannheimer C. The effect of transcutaneous electric nerve stimulation in patients with therapy-resistant hypertension. J Hum Hypertens 2000 Dec;14(12):795-8.

Chen L, Tang J, White PF, Sloninsky A, Wender RH, Naruse R, Kariger R. The effect of location of transcutaneous electrical nerve stimulation on postoperative opioid analgesic requirement:: acupoint versus nonacupoint stimulation. Anesth Analg 1998 Nov; 87(5):1129-34.

Wang B, Tang J, White PF, Naruse R, Sloninsky A, Kariger R, Gold J, Wender RH. Effect of the intensity of transcutaneous acupoint electrical stimulation on the postoperative analgesic requirement. Anesth Analg 1997 Aug; 85(2):406-13.

Chang FY, Chey WY, Ouyang A. Effect of transcutaneous nerve stimulation on esophageal function in normal subjects – evidence for a somatovisceral reflex. Am J Chin Med 1996; 24(2):185-92.

Ishimaru K, Kawakita K, Sakita M. Analgesic effects induced by TENS and electroacupuncture with different types of stimulating electrodes on deep tissues in human subjects. Pain 1995 Nov; 63(2):181-7.

Han JS, Chen XH, Yuan Y, Yan SC. Transcutaneous electrical nerve stimulation for treatment of spinal spasticity with the kappa opiate receptors, most probably dynorphin, in the central nervous system. Chin Med J (Engl) 1994 Jan; 107(1):6-11.

Yu C, Zhao S, Zhao X Treatment of simple obesity in children with photo-acupuncture. Zhongguo Zhong Xi Yi Jie He Za Zhi 1998 Jun;18(6):348-50.

Lu DP, Lu GP, Reed JF 3rd. Acupuncture/acupressure to treat gagging dental patients: a clinical study of anti-gagging effects. Gen Dent 2000 Jul-Aug; 48(4):446-52.

Norheim AJ, Pedersen EJ, Fonnebo V, Berge L. Acupressure treatment of morning sickness in pregnancy. A randomised, double-blind, placebo-controlled study. Scand J Prim Health Care. 2001 Mar; 19(1):43-7.

Werntoft E, Dykes AK. Effect of acupressure on nausea and vomiting during pregnancy. A randomized, placebo-controlled, pilot study. J Reprod Med 2001 Sep; 46(9):835-9.

Stern RM, Jokerst MD, Muth ER, Hollis C. Acupressure relieves the symptoms of motion sickness and reduces abnormal gastric activity. Altern Ther Health Med 2001 Jul-Aug; 7(4):91-4.

Windle PE, Borromeo A, Robles H, Ilacio-Uy V. The effects of acupressure on the incidence of postoperative nausea and vomiting in postsurgical patients. J Perianesth Nurs 2001 Jun; 16(3):158-62

Schlager A, Boehler M, Puhringer F. Korean hand acupressure reduces postoperative vomiting in children after strabismus surgery. Br J Anaesth 2000 Aug; 85(2):267-70.

Felhendler D, Lisander B. Effects of non-invasive stimulation of acupoints on the cardiovascular system. Complement Ther Med 1999 Dec; 7(4):231-4.

Dibble SL, Chapman J, Mack KA, Shih AS. Acupressure for nausea: results of a pilot study: Oncol Nurs Forum 2000 Jan-Feb; 27(1):41-7.

Chen ML, Lin LC, Wu SC, Lin JG. The effectiveness of acupressure in improving the quality of sleep of institutionalized residents. J Gerontol A Biol Sci Med Sci 1999 Aug; 54(8):M389-94.

Hu Y, Zhang W, Chen L. The effect of Vit. B12 injection into acupoints in the treatment of verruca plana. Zhonghua Kou Qiang Yi Xue Za Zhi 1998 Mar; 33(2):122-3.

Kwon YB, Kang MS, Han HJ, Beitz AJ, Lee JH. Visceral antinociception produced by bee venom stimulation of the Zhongwan acupuncture point in mice: role of alpha(2) adrenoceptors. Neurosci Lett 2001 Aug 3;308(2):133-7.

Kwon YB, Kang MS, Kim HW, Ham TW, Yim YK, Jeong SH, Park DS, Choi DY, Han HJ, Beitz AJ, Lee JH. Antinociceptive effects of bee venom acupuncture (apipuncture) in rodent animal models: a comparative study of acupoint versus non-acupoint stimulation. Acupunct Electrother Res 2001;26(1-2):59-68.

Kwon YB, Lee JD, Lee HJ, Han HJ, Mar WC, Kang SK, Beitz AJ, Lee JH. Bee venom injection into an acupuncture point reduces arthritis associated edema and nociceptive responses. Pain 2001 Feb 15; 90(3):271-80.

Jiang Y, Chen Y. Treatment of biliary colic by water injection in the region of Qimen, Riyue, and Juque points. J Tradit Chin Med 1995 Sep; 15(3):185-8.

Wu CC, Chen MF, Lin CC. Absorption of subcutaneous injection of Tc-99m pertechnetate via acupuncture points and non-acupuncture points. Am J Chin Med 1994;22(2):111-8.

Chen MF, Wu CC, Jong SB, Lin CC. Radionuclide venography by subcutaneous injection of Tc-99m pertechnetate at acupuncture point K-3: a case report. Am J Chin Med 1994; 22(3-4):337-40.

McMillan AS, Blasberg B. Pain-pressure threshold in painful jaw muscles following trigger point injection. J Orofac Pain 1994 Fall; 8(4):384-90.

Liu X, Sun L, Xiao J, Yin S, Liu C, Li Q, Li H, Jin B. Effect of acupuncture and point-injection treatment on immunologic function in rheumatoid arthritis. J Tradit Chin Med 1993 Sep; 13(3):174-8.

Yang LC, Jawan B, Chen CN, Ho RT, Chang KA, Lee JH. Comparison of P6 acupoint injection with 50% glucose in water and intravenous droperidol for prevention of vomiting after gynecological laparoscopy. Acta Anaesthesiol Scand 1993 Feb; 37(2): 192-4.

Wu CC, Jong SB. Radionuclide venography of lower limbs by subcutaneous injection: a clinical evaluation. Ann Nucl Med 1993 Feb; 7(1):11-9.

Chen MF, Wu CC, Jong SB, Lin CC. Differences in acupuncture point SP-10 and non-acupuncture point following subcutaneous injection of Tc-99m pertechnetate. Am J Chin Med 1993; 21(3-4):221-9.

Chen Y. Clinical research on treating senile dementia by combining acupuncture with acupoint-injection. Acupunct Electrother Res 1992; 17(2):61-73.

Zhou RL, Zhang JC. Desensitive treatment with positive allergens in acupoints of the head for allergic rhinitis and its mechanism. Zhong Xi Yi Jie He Za Zhi 1991 Dec; 11(12):721-3, 708.

Wu CC, Jong SB, Lin CC, Chen MF, Chen JR, Chung C. Subcutaneous injection of 99mTc pertechnetate at acupuncture points K-3 and B-60. Radioisotopes 1990 Jun; 39(6):261-3.

Wu CC, Jong SB. Radionuclide venography of lower limbs by subcutaneous injection: comparison with venography by intravenous injection. Ann Nucl Med 1989 Nov; 3(3):125-33.

Deng Y, Zeng T, Zhou Y, Guan X. The influence of electroacupuncture on the mast cells in the acupoints of the stomach meridian. Zhen Ci Yan Jiu 1996; 21(3):68-70.

Deng Y, Fu Z, Dong H, Wu Q, Guan X. Effects of electroacupuncture on the subcutaneous mast cells of zusanli acupoint in rat with unilateral sciatic nerve transection. Zhen Ci Yan Jiu 1996; 21(3):46-9.

Kimura M, Tohya K, Kuroiwa K, Oda H, Gorawski EC, Hua ZX, Toda S, Ohnishi M, Noguchi E. Electron microscopical and immunohistochemical studies on the induction of "Qi" employing needling manipulation. Am J Chin Med 1992; 20(1): 25-35.

Tang Z, Song X, Li J, Hou Z, Xu. S Studies on anti-inflammatory and immune effects of moxibustion.. Zhen Ci Yan Jiu 1996; 21(2):67-70.

Chiba A, Nakanishi H, Chichibu S. Thermal and antiradical properties of indirect moxibustion. Am J Chin Med 1997; 25(3-4):281-7.

Nakayama R, Oda M, Satouchi K, Saito K. Generation of acetyl glyceryl ether phosphorylcholine from the rat skin and muscle tissues stimulated by moxibustion. Biochem Biophys Res Commun 1985 Mar 15;127(2):629-34.

Wu HG, Zhou LB, Pan YY, Huang C, Chen HP, Shi Z, Hua XG. Study of the mechanisms of acupuncture and moxibustion treatment for ulcerative colitis rats in view of the gene expression of cytokines. World J Gastroenterol 1999 Dec; 5(6):515-517.

Wu P, Cao Y, Wu J. Effects of moxa-cone moxibustion at Guanyuan on erythrocytic immunity and its regulative function in tumor-bearing mice. J Tradit Chin Med 2001 Mar; 21(1):68-71.

Matsumoto T, Terasawa S. Influence of acupuncture and moxibustion on QOL of the elderly living in nursing home and care house. Nippon Ronen Igakkai Zasshi 2001 Mar; 38(2):205-11.

Murase K, Kawakita K. Diffuse noxious inhibitory controls in anti-nociception produced by acupuncture and moxibustion on trigeminal caudalis neurons in rats. Jpn J Physiol 2000 Feb; 50(1):133-40.

Yang C, Yan H. Observation of the efficacy of acupuncture and moxibustion in 62 cases of chronic colitis. J Tradit Chin Med 1999 Jun;19(2):111-4.

Liu Z, Sun F, Li J, Han Y, Wei Q, Liu C. Application of acupuncture and moxibustion for keeping shape. J Tradit Chin Med 1998 Dec;18(4):265-71.

Zheng H, Wang S, Shang J, Chen G, Huang C, Hong H, Chen S. Study on acupuncture and moxibustion therapy for female urethral syndrome. J Tradit Chin Med 1998 Jun; 18(2):122-7.

Zhai D, Din B, Liu R, Hua X, Chen H. Regulation on ACTH, beta-EP and immune function by moxibustion on different acupoints. Zhen Ci Yan Jiu 1996; 21(2):77-81.

Lee HS, Yu YC, Kim ST, Kim KS. Effects of moxibustion on blood pressure and renal function in spontaneously hypertensive rats. Am J Chin Med 1997; 25(1):21-6.

Zhang S, Chen H, Gui J, Xu C, Zhu P, Cao Z. Clinical and experimental researches in the inhibition of bile pigment lithogenesis by acupuncture and moxibustion. Zhen Ci Yan Jiu 1995; 20(3):40-5.

Iwa M, Sakita M. Effects of acupuncture and moxibustion on intestinal motility in mice. Am J Chin Med 1994; 22(2):119-25.

Zhai D, Chen H, Wang R, Hua X, Ding B, Jiang Y. Regulation on beta-END in tumor-bearing mice by moxibustion on Guanyuan point. Zhen Ci Yan Jiu 1994; 19(1):63-5, 58.

Zhang T, Gao C, Guo Y. Effects of moxibustion on the function of MDR gene product, P-glycoprotein (P-170). Zhen Ci Yan Jiu 1994; 19(2):69-71.

Fang ZR, Li YH. The observation on analgesic effect of moxibustion in rats. Zhen Ci Yan Jiu 1993; 18(4):296-9 [Article in Chinese]

Zhu LX, Li CY, Ji CF, Yang B, Li WM. The role of substance P and somatostatin in acupuncture and moxibustion-induced postsynaptic inhibition. Zhen Ci Yan Jiu 1993; 18(4):290-5.

Yan Z, Chi Y, Wang P, Cheng J, Wang Y, Shu Q, Huang G. Studies on the luminescence of channels in rats and its law of changes with "syndromes" and treatment of acupuncture and moxibustion. J Tradit Chin Med 1992 Dec; 12(4):283-7.

Chen HL, Huang XM. Treatment of chemotherapy-induced leukocytopenia with acupuncture and moxibustion. Zhong Xi Yi Jie He Za Zhi 1991 Jun;11(6):350-2, 325.

Dr. C. Shang's theory

Shang C. Electrophysiology of growth control and acupuncture. Life Sci 2001 Feb 9; 68(12):1333-42.

Shang C. Singular point, organizing center and acupuncture point. Am J Chin Med 1989; 17(3-4):119-27.

Bozhkova VP, Rozanova NV. Current status of the gap junction problems and views on their role in development. Ontogenez 1998 Jan-Feb; 29(1):5-20.

Lo CW. Genes, gene knockouts, and mutations in the analysis of gap junctions. Dev Genet 1999; 24(1-2):1-4.

Dermietzel R. Gap junction wiring: a 'new' principle in cell-to-cell communication in the nervous system? Brain Res Rev 1998 May; 26(2-3):176-83.

Bruzzone R, Ressot C. Connexins, gap junctions and cell-cell signaling in the nervous system. Eur J Neurosci 1997 Jan; 9(1):1-6.

Bibliography for further reading on water clusters

Shui-yin Lo, Anomalous state of Ice. Modern Physics Lett B, V 10(1996) 909-919.

Shui-yin Lo, A. Lo, W.C. Li, Li T.H., Li H. H., and Xu Geng. Physical properties of water with I E structures. Modern Physics Lett. B, v 10 (1996) 921-930.

Shui-yin Lo and B Bonavida: Proceedings of First Int. Sypm. of Physical, Chemical, and Biological Properties of Stable Water (I E) Clusters. World Scientific, 1998. Three papers, pp3-80.

Shui-yin Lo and Wenchong Li: Onsager's formula, conductivity, and possible new Phase Transition. Modern Physics Lett. B, V .13 1999; 885-893.

Shui-yin Lo and W. C. Li and S. H. Huang. Water clusters in life. Medical Hypotheses 2000; v54(6), 948-953.

Shui-yin Lo, Meridians in acupuncture and infrared imaging, Medical Hypotheses 2002; v58(1),72-76.

Chapter 2.

Zhang XH, Yang KN, Li XM, Ma WB. The effects of moxibustion at "zusanli" point on serum endocrine hormones of rats. Zhen Ci Yan Jiu 1989; 14(4):442-5.

Liu XA, Jiang MC, Huang PB, Zou T. Role of afferent C fibers in electroacupuncture of "zusanli" point in activating nucleus raphe magnus. Sheng Li Xue Bao 1990 Dec; 42(6): 523-33.

Wu H, Chen X. Effect of electro-acupuncture of zusanli on unit discharges in the lateral hypothalamic area induced by stomach distension. Zhen Ci Yan Jiu 1990; 15(3):194-6.

Yang J, Liu WY, Song CY. Effect of acupuncture of zusanli (ST36) on the content of beta-endorphin of the gastrointestinal tract in rats. Zhong Xi Yi Jie He Za Zhi 1989 Nov; 9(11):677-8, 646.

Jin YX, Fu Q, Guo XQ. Effects of electroacupuncture of "zusanli" acupoint on high blood pressure and blood hyperviscosity in stress rats. J Tongji Med Univ 1992; 12(4):209-15.

Yu Y, Kasahara T, Sato T, Guo SY, Liu Y, Asano K, Hisamitsu T. Enhancement of splenic interferon-gamma, interleukin-2, and NK cytotoxicity by S36 acupoint acupuncture in F344 rats. Jpn J Physiol 1997 Apr; 47(2):173-8.

Deng Y, Fu Z, Dong H, Wu Q, Guan X. Effects of electroacupuncture on the subcutaneous mast cells of zusanli acupoint in rat with unilateral sciatic nerve transection. Zhen Ci Yan Jiu 1996; 21(3):46-9.

Meng Z. The functional connection among the "zusanli"-spinal dorsal horn neurons-trigeminal sensory nucleus of rats. Zhen Ci Yan Jiu 1995; 20(3):29-32.

Liu HJ, Hsu SF, Hsieh CC, Ho TY, Hsieh CL, Tsai CC, Lin JG. The effectiveness of Tsu-San-Li (ST36) and Tai-Chung (LI3) acupoints for treatment of acute liver damage in rats. Am J Chin Med 2001; 29(2):221-6.

Xiong K, Li H, Wang T. Origin of nitric oxide synthase positive nerve fibers at zusanli area in rats. Zhongguo Zhong Xi Yi Jie He Za Zhi 1998 Apr; 18(4):230-2.

Wang Z, Huang W, Xu Q, Huang K, Cai H, Zhang X. The effect of electro-acupuncture on the adrenal gland of endotoxic shocked rats. Zhen Ci Yan Jiu 1996; 21(1):73-5.

Plant meridians

Hou TZ, Dawitof M, Wang JY, Li MD. Am J Chin Med 94; 22(1):1-10 Experimental evidence of a plant meridian system: I. Bioelectricity and acupuncture effects on electrical resistance of the soybean (Glycine max).

Hou TZ, Re ZW, Li MD. Am J Chin Med 1994; 22(2):103-10 Experimental evidence of a plant meridian system: II. The effects of needle acupuncture on the temperature changes of soybean (Glycine max).

Hou TZ, Luan JY, Wang JY, Li MD. Am J Chin Med 1994; 22(3-4):205-14. Experimental evidence of a plant meridian system: III. The sound characteristics of phylodendron (Alocasia) and effects of acupuncture on those properties.

Hou TZ, Mooneyham RE. Am J Chin Med 1999; 27(1):1-10 Applied studies of plant meridian system: I. The effect of agri-wave technology on yield and quality of tomatos.

Hou TZ, Li MD. Am J Chin Med 1997; 25(3-4):253-61. Experimental evidence of a plant meridian system: V. Acupuncture effect on circumnutation movements of shoots of Phaselus vulgaris L. pole bean.

Hou TZ, Li MD. Am J Chin Med 1997; 25(2):135-42. Experimental evidence of a plant meridian system: IV. The effects of acupuncture on growth and metabolism of Phaseolus vulgaris L. beans.

Chapter 3.

A. Immune system

Gollub RL, Hui KK, Stefano GB. Acupuncture: pain management coupled to immune stimulation. Zhongguo Yao Li Xue Bao 1999 Sep;20 (9):769-77.

Joos S, Schott C, Zou H, Daniel V, Martin E. Immunomodulatory effects of acupuncture in the treatment of allergic asthma: a randomized controlled study. J Altern Complement Med 2000 Dec; 6(6):519-25.

Miller AL. The etiologies, pathophysiology, and alternative/complementary treatment of asthma. Altern Med Rev 2001 Feb; 6(1):20-47.

Sun T, Du LN, Wu GC, Cao XD. Effect of intrathecal morphine and electro-acupuncture on cellular immune function of rats and increment of mu-opioid receptor mRNA expression in PAG following intrathecal morphine. Acupunct Electrother Res 2000; 25(1):1-8.

Nepp J, Derbolav A, Haslinger-Akramian J, Mudrich C, Schauersberger J, Wedrich A. Effect of acupuncture in keratoconjunctivitis sicca. Klin Monatsbl Augenheilkd 1999 Oct; 215(4):228-32.

Petti F, Bangrazi A, Liguori A, Reale G, Ippoliti F. Effects of acupuncture on immune response related to opioid-like peptides. J Tradit Chin Med 1998 Mar; 18(1):55-63.

B. Nervous system

Chen YS, Yao CH, Chen TH, Lin JG, Hsieh CL, Lin CC, Lao CJ, Tsai CC. Effect of acupuncture stimulation on peripheral nerve regeneration using silicone rubber chambers. Am J Chin Med 2001; 29(3-4):377-85.

Yu Q, Shen PQ, Li XH. Experimental study on functional rehabil-itation of peripheral nerve with electric acupuncture. Zhongguo Xiu Fu Chong Jian Wai Ke Za Zhi 2001 Sep; 15(5):315-7.

Stener-Victorin E, Lundeberg T, Waldenstrom U, Manni L, Aloe L, Gunnarsson S, Janson PO. Effects of electro-acupuncture on nerve growth factor and ovarian morphology in rats with experimentally induced polycystic ovaries. Biol Reprod 2000 Nov; 63(5): 1497-503.

Haker E, Egekvist H, Bjerring P. Effect of sensory stimulation (acupuncture) on sympathetic and parasympathetic activities in healthy subjects. J Auton Nerv Syst 2000 Feb 14; 79(1):52-9.

Knardahl S, Elam M, Olausson B, Wallin BG. Sympathetic nerve activity after acupuncture in humans. Pain 1998 Mar; 75(1):19-25.

Okada K, Oshima M, Kawakita K. Examination of the afferent fiber responsible for the suppression of jaw-opening reflex in heat, cold, and manual acupuncture stimulation in rats. Brain Res 1996 Nov 18; 740(1-2):201-7.

Hao J, Zhao C, Cao S, Yang S. Electric acupuncture treatment of peripheral nerve injury. J Tradit Chin Med 1995 Jun; 15(2):114-7.

C. Reproductive system

Zhu D, Ma Q, Li C, Wang L. Effect of stimulation of shenshu point on the aging process of genital system in aged female rats and the role of monoamine neurotransmitters. J Tradit Chin Med 2000 Mar; 20(1):59-62.

Stener-Victorin E, Waldenstrom U, Tagnfors U, Lundeberg T, Lindstedt G, Janson PO. Effects of electro-acupuncture on anovulation in women with polycystic ovary syndrome. Acta Obstet Gynecol Scand 2000 Mar; 79(3):180-8.

Kho HG, Sweep CG, Chen X, Rabsztyn PR, Meuleman EJ. The use of acupuncture in the treatment of erectile dysfunction. Int J Impot Res 1999 Feb; 11(1):41-6.

Chen BY. Acupuncture normalizes dysfunction of hypothalamic-pituitary-ovarian axis. Acupunct Electrother Res 1997; 22(2):97-108.

Chen BY, Cheng LH, Gao H, Ji SZ. Effects of electroacupuncture on the expression of estrogen receptor protein and mRNA in rat brain. Sheng Li Xue Bao 1998 Oct; 50(5):495-500.

Guo C, Zhang W, Zheng S, Ju D, Zhao C. Clinical observation on efficacy of electro-acupuncture therapy in hyperplasia of mammary glands and its effect on immunological function. J Tradit Chin Med 1996 Dec; 16(4):281-7.

Stener-Victorin E, Waldenstrom U, Andersson SA, Wikland M. Reduction of blood flow impedance in the uterine arteries of infertile women with electro-acupuncture. Hum Reprod 1996 Jun; 11(6):1314-7.

Chen B, Ji S, Gao H, He L. The effects of electroacupuncture treatment on nucleolar organizer regions of adrenal cortex in ovariectomized rats. Zhen Ci Yan Jiu 1994; 19(1):46-50.

Kong T, Fan T, Chu X. Studies on the relationship between acupuncture analgesia and testosterone or dihydrotestosterone in blood plasma. Zhen Ci Yan Jiu 1991; 16(2): 138-41.

Li Q, Wang L. Clinical observation on correcting malposition of fetus by electro-acupuncture. J Tradit Chin Med 1996 Dec; 16(4):260-2.

Yang SP, Yu J, He L. Release of gonadotropin-releasing hormone (GnRH) from the medio-basal hypothalamus induced by electroacupuncture in conscious female rabbits. Acupunct Electrother Res 1994 Jan-Mar; 19(1):19-27.

Chen BY, Yu J. Relationship between blood radioimmunoreactive beta-endorphin and hand skin temperature during the electro-acupuncture induction of ovulation. Acupunct Electrother Res 1991; 16(1-2):1-5.

Parshutin NP, Korsakov SG. Comparative analysis of the data of acupuncture electrodiagnosis and hormonal status of women with oligomenorrhea. Akush Ginekol (Mosk) 1990 Jun; (6):26-9.

Tsuei JJ, Lai Y, Sharma SD. The influence of acupuncture stimulation during pregnancy: the induction and inhibition of labor. Obstet Gynecol 1977 Oct; 50(4):479-8.

Lin Y, Zhou Z, Shen W, Shen J, Hu M, Zhang F, Hu P, Xu M, Huang S, Zheng Y. Clinical and experimental studies on shallow needling technique for treating childhood diarrhea. J Tradit Chin Med 1993 Jun; 13(2):107-14.

Hwang YC, Jenkins EM. Effect of acupuncture on young pigs with induced enteropathogenic Escherichia coli diarrhea. Am J Vet Res 1988 Sep; 49(9):1641-3.

Broide E, Pintov S, Portnoy S, Barg J, Klinowski E, Scapa E. Effectiveness of acupuncture for treatment of childhood constipation. Dig Dis Sci 2001 Jun; 46(6):1270-5.

Li Y, Tougas G, Chiverton SG, Hunt RH. The effect of acupuncture on gastrointestinal function and disorders. Am J Gastroenterol 1992 Oct; 87(10):1372-81.

Stern RM, Jokerst MD, Muth ER, Hollis C. Acupressure relieves the symptoms of motion sickness and reduces abnormal gastric activity. Altern Ther Health Med 2001 Jul-Aug; 7(4):91-4.

Wan Q. Auricular-plaster therapy plus acupuncture at zusanli for postoperative recovery of intestinal function. J Tradit Chin Med 2000 Jun; 20(2):134-5.

Xu F, Chen R.. Reciprocal actions of acupoints on gastrointestinal peristalsis during electroacupuncture in mice. J Tradit Chin Med 1999 Jun; 19(2):141-4.

Chen R, Kang M. Observation on frequency spectrum of electrogastrogram (EGG) in acupuncture treatment of functional dyspepsia. J Tradit Chin Med 1998 Sep; 18(3): 184-7.

Zhang X, Yuan Y, Kuang P, Wu W, Zhang F, Liu J. The changes of vasoactive intestinal peptide somatostatin and pancreatic polypeptide in blood and CSF of acute cerebral infarction patients and the effect of acupuncture on them. Zhen Ci Yan Jiu 1996; 21(4):10-6.

Chan J, Carr I, Mayberry JF. The role of acupuncture in the treatment of irritable bowel syndrome: a pilot study. Hepatogastroenterology 1997 Sep-Oct; 44(17):1328-30.

Borozan S, Petkovic G. Ear acupuncture has a hypotonic effect on the gastrointestinal tract. Vojnosanit Pregl 1996 Jan-Feb; 53(1):31-3.

Iwa M, Sakita M. Effects of acupuncture and moxibustion on intestinal motility in mice. Am J Chin Med 1994; 22(2):119-25.

Xu G. Regulating effect of electro-acupuncture on dysrythmia of gastro-colonic electric activity induced by erythromycine in rabbits. Zhen Ci Yan Jiu 1994; 19(1):71-4.

Xu G. Influence of stress on gastroenteric electric activity and modulated effect of acupuncture on it in rats. Zhen Ci Yan Jiu 1994; 19(2):72-4.

Ma C, Liu Z. Regulative effects of electroacupuncture on gastric hyperfunction induced by electrostimulation of the lateral hypothalamus area of rabbits. Zhen Ci Yan Jiu 1994; 19(2):42-6.

Sato A, Sato Y, Suzuki A, Uchida S. Neural mechanisms of the reflex inhibition and excitation of gastric motility elicited by acupuncture-like stimulation in anesthetized rats. Neurosci Res 1993 Oct; 18(1):53-62.

Xiang L, Zhu F, Ma Y, Weng E, Tang G. Influences of acupuncture on gastroduodenal mucosal lesion and electrical changing induced by stress in rats. Zhen Ci Yan Jiu 1993; 18(1):53-7. [Article in Russian]

Kopeikin VN, Belentsova LA. Acupuncture in the treatment of gastroenteric diseases in children. Vopr Kurortol Fizioter Lech Fiz Kult 1991 May-Jun; (3):43-5.

Liu JX, Zhao Q. Effect of acupuncture on intestinal motion and sero-enzyme activity in perioperation. Zhong Xi Yi Jie He Za Zhi 1991 Mar; 11(3):156-7, 133-4.

Len J, Xu G, Liu W, Zhang Q. The regulating effect of electroacupuncture on gastroenteric electric activity in guinea pigs of peripheral vomiting. Zhen Ci Yan Jiu 1991; 16(1): 69-72.

Yang J, Liu WY, Song CY. Effect of acupuncture of zusanli (ST36) on the content of beta-endorphin of the gastrointestinal tract in rats. Zhong Xi Yi Jie He Za Zhi 1989 Nov; 9(11):677-8, 646.

Kabanov AN, Vozliublennyi SI, Platonov NS, Kozhukhin AP, Bronfina ZL. Use of acupuncture for restoring motor and transit functions of the stomach and intestines in suppurative peritonitis. Klin Khir 1989; (1):33-4.

Kapustin AV, Khavkin AI. State of the autonomic nervous system in children with disorders of the motor-evacuatory function of the upper part of the gastrointestinal tract. Pediatriia 1989; (1):68-71.

D. Urinary system

Lin TB, Fu TC, Chen CF, Lin YJ, Chien CT. Low and high frequency electro-acupuncture at Hoku elicits a distinct mechanism to activate sympathetic nervous system in anesthetized rats. Neurosci Lett 1998 May 15; 247(2-3):155-8.

Lee HS, Yu YC, Kim ST, Kim KS. Effects of moxibustion on blood pressure and renal function in spontaneously hypertensive rats. Am J Chin Med 1997; 25(1):21-6.

Kitakoji H, Terasaki T, Honjo H, Odahara Y, Ukimura O, Kojima M, Watanabe H. Effect of acupuncture on the overactive bladder. Nippon Hinyokika Gakkai Zasshi 1995 Oct; 86(10):1514-9

Ben H, Zhu Y. The effect of electro-acupuncture at auricular & body acupoints on the curve changes of the pressure-volume of urinary bladder and the electric activity of pelvic nerves of rat. Zhen Ci Yan Jiu 1995; 20(2):51-4.

Xu N, Xu G, Zhu C. Effect of electroacupuncture at "shenshu" point on renal blood flow in rabbits. Zhen Ci Yan Jiu 1995; 20(2):48-50.

Kachan AT, Trubin MIu, Skoromets AA, Shmushkevich AI. Acupuncture reflexotherapy of neurogenic bladder dysfunction in children with enuresis. Zh Nevropatol Psikhiatr Im S S Korsakova 1993; 93(5):40-2.

Yao T. Acupuncture and somatic nerve stimulation: mechanism underlying effects on cardiovascular and renal activities. Scand J Rehabil Med Suppl 1993; 29:7-18.

Ben H, Zu Y, Ye Y. The effect of electroacupuncture on the function of the partially denervated bladder in rabbits. Zhen Ci Yan Jiu 1993; 18(1):68-72.

Xu N. Effect of electroacupuncture at "taixi" point on plasma thromboxane A2 and prostacyclin in the rabbit with renal ischemia. Zhen Ci Yan Jiu 1993; 18(3):240-2.

Sato A, Sato Y, Suzuki A. Mechanism of the reflex inhibition of micturition contractions of the urinary bladder elicited by acupuncture-like stimulation in anesthetized rats. Neurosci Res 1992 Nov; 15(3):189-98.

E. Heart

Syuu Y. Matsubara H. Kiyooka T. Hosoqi S. Mohri S, Araki J, Ohe T, Suga H. Cardiovascular beneficial effects of electroacupuncture at Neiguan (PC-6) acupoint in anesthetized open-chest dog. Jpn J Physiol. 2001 Apr; 51(2):231-8.

Cao Q, Liu J, Chen S, Han Z. Effects of electroacupuncture at neiguan on myocardial microcirculation in rabbits with acute myocardial ischemia. J Tradit Chin Med. 1998 Jun; 18(2):134-9.

Chao DM, Shen LL, Tien-A-Looi S, Pitsillides KF, Li P, Longhurst JC. Naloxone reverses inhibitory effect of electroacupuncture on sympathetic cardiovascular reflex responses. Am J Physiol. 1999 Jun; 276(6 Pt 2):H2127-34.

Wang W, Zhang Y, Guo Y, miao, Wang X, Xutang P, The detection and studies on the change of H+ concentration in the regular points of the rabbit suffering from arrhythmia induced by aconitine. Zhen Ci Yan Jiu. 1996; 21(4):59-63. Chinese

Liu J, Chen S, Cao Q., Zhang J. Influence of neuronal excitation and inhibition of rostral ventrolateral medulla on the effect of electroacupuncture of "Neiguan" acupoint Zhen Ci Yan Jiu. 1996; 21(1):34-8. Chinese.

Lin JG, Ho SJ, Lin JC, Effect of acupuncture on cardiopulmonary function. Chin Med J (Engl). 1996 Jun; 109(6):482-5.

Shi X, Wang ZP, Liu KX, Effect of acupuncture on heart rate variability in coronary heart disease patients. Zhongguo Zhong Xi Yi Jie He Za Zhi. 1995 Sep; 15(9):536-8. Chinese.

Li L, Chen H. Xi Y, Wang X, Han G, Zhou Y, Yang D, Zhao W, Feng Z, Jiao B et al. Comparative observation on effect of electric acupuncture of neiguan (P6) at chen time versus xu time on left ventricular function in patients with coronary heart disease. Tradit Chin Med. 1994 Dec; 14(4):262-5.

Liu J, Han Z, Chen S, Cao Q. Influence of electroacupuncture of neiguan (PC 6) on ami-induced changes in electrical activity of dorsal horn neurons. Zhen Ci Yan Jiu. 1994; 19(1):37-41. Chinese.

You Z, Hu X, Wu B, Zhang W, Liang D. Experimental observation on the relation between pericardium meridian and cardiac function. Zhen Ci Yan Jiu. 1993; 18(2): 143-8. Chinese.

Cao Q, Han Z, Chen S. Influence of EA on the effective refractory period in rabbits with AMI. Zhen Ci Yan Jiu. 1993; 18(4):276-9. Chinese.

Gao C, Meng J, Fu W, Song L. Effect of electroacupuncture on myocardial oxygen metabolism and pH of coronary sinus blood during experimental angina pectoris. Zhen Ci Yan Jiu. 1992; 17(1):28-32. Chinese.

Richter A, Herlitz J, Hialmarson A. Effect of acupuncture in patients with angina pectoris. Eur Heart J. 1991 Feb; 12(2):175-8.

Wang T, Pan Q, An experimental study on acupoint neiguan-heart short reflex. Zhen Ci Yan Jiu. 1991; 16(2):115-9. Chinese.

Gao C, Meng J, Fu W, Song L. Change of myocardial glucose and free fatty acid metabolism and effect of electroacupuncture on them during experimental myocardial angina. Zhen Ci Yan Jiu. 1990; 15(1):66-70. Chinese.

Cao Q, Liu J, Wei Y, Han Z. Studies on the interrelationships between some acupoints on pericardial channel and heart. Zhen Ci Yan Jiu. 1990; 15(1):35-9. Chinese.

Song X, Tang Z, Hou Z, Shan H, Chen Y, The anti-hemorrhagic shock of acupuncture on neiguan and its effect on the cardial pump function and the blood viscosity. Zhen Ci Yan Jiu. 1990; 15(1):30-4. Chinese.

F. Spleen

Sun T, Du LN, Wu GC, Cao. XD.Effect of intrathecal morphine and electro-acupuncture on cellular immune function of rats and increment of mu-opioid receptor mRNA expression in PAG following intrathecal morphine. Acupunct Electrother Res. 2000; 25(1):1-8.

Dong L, Yuan D, Fan L, Su L, Fu Z. Effect of HE-NE laser acupuncture on the spleen in rats. Zhen Ci Yan Jiu 1996; 21(4):64-7.

Yu Y, Kasahara T, Sato T, Guo SY, Liu Y, Asano K, Hisamitsu T. Enhancement of splenic interferon-gamma, interleukin-2, and NK cytotoxicity by S36 acupoint acupuncture in F344 rats. Jpn J Physiol. 1997 Apr; 47(2):173-8.

Cheng XD, Wu GC, He QZ, Cao XD. Effect of continued electroacupuncture on induction of interleukin-2 production of spleen lymphocytes from the injured rats. Acupunct Electrother Res 1997; 22(1):1-8.

Sato T, Yu Y, Guo SY, Kasahara T, Hisamitsu T. Acupuncture stimulation enhances splenic natural killer cell cytotoxicity in rats. Jpn J Physiol 1996 Apr; 46(2):131-6.

G. Stomach

Lee CH, Jung HS, Lee TY, Lee SR, Yuk SW, Lee KG, Lee BH. Studies of the central neural pathways to the stomach and Zusanli (ST36). Am J Chin Med 2001; 29(2):211-20.

Hu S, Guo Y, Zhang Y, Wang Y, Xu T. Effect of acupuncture on the restrained state of stomach after injecting TFP in zusanli point. Zhen Ci Yan Jiu 1996; 21(2):57-61.

Sun Y, Xu GS, Liu WP, Xu NG. The role of NO/ET and the effect of electro-acupuncture on injured gastric mucosa in rats. Sheng Li Xue Bao 1999 Apr; 51(2):206-10.

Deng Y, Zeng T, Zhou Y, Guan X. The influence of electroacupuncture on the mast cells in the acupoints of the stomach meridian. Zhen Ci Yan Jiu 1996; 21(3):68-70.

Noguchi E, Hayashi H. Increases in gastric acidity in response to electroacupuncture stimulation of the hindlimb of anesthetized rats. Jpn J Physiol 1996 Feb; 46(1):53-8.

Shen D, Wei D, Liu B, Zhang F. Effects of electroacupuncture on gastrin, mast cell and gastric mucosal barrier in the course of protecting rat stress peptic ulcer. Zhen Ci Yan Jiu 1995; 20(3):46-9.

Fu Z. A study on factors of acupuncture methods affecting effects through acupuncture treatments for experimental gastric ulcer in rats. Zhen Ci Yan Jiu 1995; 20(2):40-4.

Liu X, Zhang S, Wei B. The relation of the relative specificity of point to channel lines or spinal segments. Zhen Ci Yan Jiu 1995; 20(1):54-9.

Shen D, Liu B, Wi D, Zhang F, Chen Y. Effects of electroacupuncture on central and peripheral monoamine neurotransmitter in the course of protecting rat stress peptic ulcer. Zhen Ci Yan Jiu 1994; 19(1):51-4.

Wu H, Chen X. Effect of electro-acupuncture of zusanli on unit discharges in the lateral hypothalamic area induced by stomach distension. Zhen Ci Yan Jiu 1990; 15(3):194-6.

Pan C, Jin W, Shen D. An observation of protective effect of acupuncture on the gastric mucosa of Wistar rats and the relative histochemical changes of the neurotransmitters. Zhen Ci Yan Jiu 1990; 15(1):48-54.

Andreev BV, Vasil'ev IuN, Ignatov IuD, Kachan AT, Bogdanov NN. Effect of electroacupuncture on signs of emotional stress caused by pain. : Biull Eksp Biol Med 1981; 91(1):18-20.

Lin X, Liang J, Ren J, Mu F, Zhang M, Chen JD. Electrical stimulation of acupuncture points enhances gastric myoelectrical activity in humans. Am J Gastroenterol 1997 Sep; 92(9):1527-30.

Lin X, Liang J, Ren J, Mu F, Zhang M, Chen JD. Electrical stimulation of acupuncture points enhances gastric myoelectrical activity in humans. Am J Gastroenterol 1997 Sep; 92(9):1527-30.

H. Lung

Zamotaev IP, Mamontova LI, Zavolovskaia LI, Rudakova OM. Effect of laser acupuncture on the pulmonary vascular resistance in patients with obstructive chronic lung diseases. Klin Med (Mosk) 1991 May; 69(5):68-71.

Zamotaev IP, Mamontova LI, Zavolovskaia LI. Effectiveness of using semiconductor laser in the complex treatment of patients with obstructive forms of chronic nonspecific lung diseases. Klin Med (Mosk) 1990 Jan; 68(1):66-9.

Davis CL, Lewith GT, Broomfield J, Prescott P. A pilot project to assess the methodological issues involved in evaluating acupuncture as a treatment for disabling breathlessness. J Altern Complement Med 2001 Dec; 7(6):633-9.

Lai X, Li Y, Fan Z, Zhang J, Liu B. An analysis of therapeutic effect of drug acupoint application in 209 cases of allergic asthma. J Tradit Chin Med 2001 Jun;.21(2):122-6.

Hu J. Clinical observation on 25 cases of hormone dependent bronchial asthma treated by acupuncture. J Tradit Chin Med 1998 Mar; 18(1):27-30.

Neumeister W, Kuhlemann H, Bauer T, Krause S, Schultze-Werninghaus G, Rasche K. Effect of acupuncture on quality of life, mouth occlusion pressures and lung function in COPD. Med Klin 1999 Apr; 94(1 Spec No):106-9.

Chen L, Liu X, Wang X, Yan G, Hao X, Wang L, Mu Y. Effects of ear acupuncture on beta-adrenoreceptor in lung tissues of guinea-pigs with experimental asthma. Zhen Ci Yan Jiu 1996; 21(1):56-9.

Tashkin DP, Kroening RJ, Bresler DE, Simmons M, Coulson AH, Kerschnar H. A controlled trial of real and simulated acupuncture in the management of chronic asthma. J Allergy Clin Immunol 1985 Dec; 76(6):855-64.

Jobst K, Chen JH, McPherson K, Arrowsmith J, Brown V, Efthimiou J, Fletcher HJ, Maciocia G, Mole P, Shifrin K, et al. Controlled trial of acupuncture for disabling breathlessness. Lancet 1986 Dec 20-27; 2(8521-22):1416-9.

Sliwinski J, Matusiewicz R. The effect of acupuncture on the clinical state of patients suffering from chronic spastic bronchitis and undergoing long-term treatment with corticosteroids. Acupunct Electrother Res 1984; 9(4):203-15.

Takishima T, Mue S, Tamura G, Ishihara T, Watanabe K. The bronchodilating effect of acupuncture in patients with acute asthma. Ann Allergy 1982 Jan; 48(1):44-9.

Sovijarvi AR, Poppius H. Acute bronchodilating effect of transcutaneous nerve stimulation in asthma. A peripheral reflex or psychogenic response. Scand J Respir Dis 1977 Jun; 58(3):164-9.

I. Liver

Liu HJ, Hsu The effectiveness of Tsu-San-Li (ST36) and Tai-Chung (Li-3) acupoints for treatment of acute liver damage in rats. Am J Chin Med 2001; 29(2):221-6.

Lin JG, Yang SH, Tsai CH. Acupuncture protection against experimental hyperbilirubinemia and cholangitis in rats. Am J Chin Med 1995; 23(2):131-7.

Yang J, Zhao R, Yuan J, Chen G, Zhang L, Yu M, Lu A, Zhang Z. The experimental study of prevention and treatment of the side-effects of chemotherapy with acupuncture (comparison among the effects of acupuncture at different acupoints). Zhen Ci Yan Jiu 1994; 19(1):75-8.

Chakrabarti AK, Chatterjee K, Ghosh JJ, Ganguly A. Electroacupuncture and its effect on rat hepatic functions. Acupunct Electrother Res 1983; 8(2):111-26.

Chapter 4.

A. Obesity

Liu Z, Sun F, Han Y. Effect of acupuncture on level of monoamines and activity of adenosine triphosphatase in lateral hypothalamic area of obese rats. Zhongguo Zhong Xi Yi Jie He Za Zhi 2000 Jul; 20(7):521-3.

Yu C, Zhao S, Zhao X. Treatment of simple obesity in children with photo-acupuncture. Zhongguo Zhong Xi Yi Jie He Za Zhi 1998 Jun; 18(6):348-50.

Zhao M, Liu Z, Su J. The time-effect relationship of central action in acupuncture treatment for weight reduction. Tradit Chin Med 2000 Mar; 20(1):26-9.

Liu Z, Sun F, Li J, Han Y, Wei Q, Liu C. Application of acupuncture and moxibustion for keeping shape. J Tradit Chin Med 1998 Dec; 18(4):265-71.

Li J. Clinical experience in acupuncture treatment of obesity. J Tradit Chin Med. 1999 Mar; 19(1):48-51.

Richards D, Marley J. Stimulation of auricular acupuncture points in weight loss. Aust Fam Physician 1998 Jul; 27 Suppl 2:S73-7.

Liu Z. Effects of acupuncture on lipid, TXB2, 6-keto-PGF, alpha in simple obesity patients complicated with hyperlipidemia. Zhen Ci Yan Jiu 1996; 21(4):17-21.

Sun F. The antiobesity effect of acupuncture and its influence on water and salt metabolism. Zhen Ci Yan Jiu 1996; 21(2):19-24.

Huang MH, Yang RC, Hu SH. Preliminary results of triple therapy for obesity. Int J Obes Relat Metab Disord 1996 Sep; 20(9):830-6.

Shiraishi T, Onoe M, Kageyama T, Sameshima Y, Kojima T, Konishi S, Yoshimatsu H, Sakata T. Effects of auricular acupuncture stimulation on nonobese, healthy volunteer subjects. Obes Res 1995 Dec; 3 Suppl 5:667S-673S.

Liu ZC, Sun FM, Wang YZ. Good regulation of acupuncture in simple obesity patients with stomach-intestine excessive heat type. Zhongguo Zhong Xi Yi Jie He Za Zhi 1995 Mar; 15(3):137-40.

Shiraishi T, Onoe M, Kojima T, Sameshima Y, Kageyama T. Effects of auricular stimulation on feeding-related hypothalamic neuronal activity in normal and obese rats. Brain Res Bull 1995; 36(2):141-8.

Shafshak TS. Electroacupuncture and exercise in body weight reduction and their application in rehabilitating patients with knee osteoarthritis. Am J Chin Med 1995; 23(1):15-25.

Liu Z, Sun F, Li J, Wang Y, Hu K. Effect of acupuncture on weight loss evaluated by adrenal function. J Tradit Chin Med 1993 Sep; 13(3):169-73.

Asamoto S, Takeshige C. Activation of the satiety center by auricular acupuncture point stimulation. Brain Res Bull 1992 Aug; 29(2):157-64.

Liu Z, Sun F, Li J, Shi X, Hu L, Wang Y, Qian Z. Prophylactic and therapeutic effects of acupuncture on simple obesity complicated by cardiovascular diseases. J Tradit Chin Med 1992 Mar; 12(1):21-9.

Liu ZC, Sun FM, Shen DZ. Effect of acupuncture and moxibustion on antiobesity in the variation of plasma cyclic nucleotide and the function of vegetative nervous system. Zhong Xi Yi Jie He Za Zhi 1991 Feb; 11(2):83-6, 67-8.

Liu ZC. Effect of acupuncture and moxibustion on hypothalamus-pituitary-adrenal axis suffering from simple obesity. Zhong Xi Yi Jie He Za Zhi 1990 Nov; 10(11):656-9, 643-4.

Liu ZC. Regulatory effects of acupuncture and moxibustion on simple obesity complicated with hypertension. Zhong Xi Yi Jie He Za Zhi 1990 Sep; 10(9):522-5, 515.

Liu Z. Effect of acupuncture and moxibustion on the high density lipoprotein cholesterol in simple obesity. Zhen Ci Yan Jiu 1990; 15(3):227-31.

Malygin KIu, Kichev GS. The efficacy of combined general anesthesia with electroacupuncture analgesia in abdominal surgery in patients with concurrent massive obesity. Anesteziol Reanimatol 1989 Sep-Oct; (5):26-9.

B. Aging

Zhu D, Ma Q, Li C, Wang L. Effect of stimulation of shenshu point on the aging process of genital system in aged female rats and the role of monoamine neurotransmitters. J Tradit Chin Med 2000 Mar; 20(1):59-62.

Gaylord S. Alternative therapies and empowerment of older women. J Women Aging 1999; 11(2-3):29-47.

Toriizuka K, Okumura M, Iijima K, Haruyama K, Cyong JC. Acupuncture inhibits the decrease in brain catecholamine contents and the impairment of passive avoidance task in ovariectomized mice. Acupunct Electrother Res 1999; 24(1):45-57.

Omura Y, Shimotsura Y, Ooki M, Noguchi T. Estimation of the amount of telomere molecules in different human age groups and the telomere increasing effect of acupuncture and shiatsu on St.36, using synthesized basic units of the human telomere molecules as reference control substances for the bi-digital O-ring test resonance phenomenon. Acupunct Electrother Res 1998; 23(3-4):185-206.

Sato Y, Shibuya A, Adachi H, Kato R, Horita H, Tsukamoto T. Restoration of sexual behavior and dopaminergic neurotransmission by long term exogenous testosterone replacement in aged male rats. J Urol 1998 Oct; 160(4):1572-5.

C. Qigong

Wang CX, Xu DH, Qian YC. Effect of qigong on heart-qi deficiency and blood stasis type of hypertension and its mechanism. Zhongguo Zhong Xi Yi Jie He Za Zhi 1995 Aug; 15(8):454-8.

Yan Y, Luo Y, He T, Ye X, Tian M, Zhang S, Tang B. Effects of "he xiang zhuang gongfu" on respiratory function in healthy adults. Hua Xi Yi Ke Da Xue Xue Bao 1993 Jun; 24(2):225-7.

Zhang SX, Guo HZ, Jing BS, Liu SF. The characteristics and significance of intrathoracic and abdominal pressures during Qigong (Q-G) maneuvering. Aviat Space Environ Med 1992 Sep; 63(9):795-801.

Zhang SX, Guo HZ, Jing BS, Wang X, Zhang LM. Experimental verification of effectiveness and harmlessness of the Qigong maneuver. Aviat Space Environ Med 1991 Jan; 62(1):46-52.

Kubota Y, Sato W, Toichi M, Murai T, Okada T, Hayashi A, Sengoku A. Frontal midline theta rhythm is correlated with cardiac autonomic activities during the performance of an attention demanding meditation procedure. Brain Res Cogn Brain Res 2001 Apr; 11(2):281-7.

Jones BM. Changes in cytokine production in healthy subjects practicing Guolin Qigong: a pilot study. BMC Complement Altern Med 2001; 1(1):8.

Sancier KM. The effect of qigong on therapeutic balancing measured by electroacupuncture according to Voll (EAV): a preliminary study. Acupunct Electrother Res 1994 Jun-Sep; 19(2-3):119-27.

Wu WH, Bandilla E, Ciccone DS, Yang J, Cheng SC, Carner N, Wu Y, Shen R.. Effects of qigong on late-stage complex regional pain syndrome. Altern Ther Health Med 1999 Jan; 5(1):45-54.

Iwao M, Kajiyama S, Mori H, Oogaki K. Effects of qigong walking on diabetic patients: a pilot study. J Altern Complement Med 1999 Aug; 5(4):353-8.

Yao BS. A preliminary study on the changes of T-cell subsets in patients with aplastic anemia treated with qigong. Zhong Xi Yi Jie He Za Zhi 1989 Jun; 9(6):341-2, 324.

Ryu H, Mo HY, Mo GD, Choi BM, Jun CD, Seo CM, Kim HM, Chung HT. Delayed cutaneous hypersensitivity reactions in qigong (chun do sun bup) trainees by multitest cell mediated immunity. Am J Chin Med 1995; 23(2):139-44.

Sun FL, Yan YA. Effects of various qigong breathing patterns on variability of heart rate. Zhongguo Zhong Xi Yi Jie He Za Zhi 1992 Sep; 12(9):527-30, 516.

Zhang W, Zheng R, Zhang B, Yu W, Shen X. An observation on flash evoked cortical potentials and qigong meditation. Am J Chin Med 1993; 21(3-4):243-9.

Takeshige C, Aoki T. Effect of artificial and human external qigong on electroencephalograms in rabbit and spontaneous electrical activity of the rat pineal gland. Acupunct Electrother Res 1994 Jun-Sep; 19(2-3):89-106.

Takeshige C, Sato M. Comparisons of pain relief mechanisms between needling to the muscle, static magnetic field, external qigong and needling to the acupuncture point. Acupunct Electrother Res 1996 Apr-Jun; 21(2):119-31.

Omura Y, Beckman SL. Application of intensified (+) qi gong energy, (-) electrical field, (S) magnetic field, electrical pulses (1-2 pulses/sec), strong shiatsu massage or acupuncture on the accurate organ representation areas of the hands to improve circulation and enhance drug uptake in pathological organs: clinical applications with special emphasis on the "Chlamydia-(Lyme)-uric acid syndrome" and "Chlamydia-(cytomegalovirus) -uric acid syndrome." Acupunct Electrother Res 1995 Jan-Mar; 20(1):21-72.

Omura Y, Lin TL, Debreceni L, Losco BM, Freed S, Muteki T, Lin CH. Unique changes found on the qi gong (chi gong) master's and patient's body during qi gong treatment; their relationships to certain meridians & acupuncture points and the re-creation of therapeutic qi gong states by children & adults. Acupunct Electrother Res 1989; 14(1):61-89.

Omura Y, Losco BM, Omura AK, Takeshige C, Hisamitsu T, Shimotsuura Y, Yamamoto S, Ishikawa H, Muteki T, Nakajima H, et al. Common factors contributing to intractable pain and medical problems with insufficient drug uptake in areas to be treated, and their pathogenesis and treatment: Part I. Combined use of medication with acupuncture, qi gong energy-stored material, soft laser or electrical stimulation. : Acupunct Electrother Res 1992; 17(2):107-48.

Omura Y. Storing of qi gong energy in various materials and drugs (qi gongnization): its clinical application for treatment of pain, circulatory disturbance, bacterial or viral infections, heavy metal deposits, and related intractable medical problems by selectively enhancing circulation and drug uptake. Acupunct Electrother Res 1990; 15(2):137-57.

Zhang JZ, Li JZ, He QN. Statistical brain topographic mapping analysis for EEGs recorded during qigong state. Int J Neurosci 1988 Feb; 38(3-4):415-25.

Fukushima M, Kataoka T, Hamada C, Matsumoto M. Evidence of qigong energy and its biological effect on the enhancement of the phagocytic activity of human polymorphonuclear leukocytes. Am J Chin Med 2001; 29(1):1-16.

Liu B, Jiao J, Li Y. Effect of qigong exercise on the blood level of monoamine neurotransmitters in patients with chronic diseases. Zhong Xi Yi Jie He Za Zhi 1990 Apr; 10(4):203-5, 195.

Lee MS, Kim BG, Huh HJ, Ryu H, Lee HS, Chung HT. Effect of Qi-training on blood pressure, heart rate and respiration rate. Clin Physiol 2000 May; 20(3):173-6.

Lee MS, Kang CW, Ryu H, Kim JD, Chung HT. Effects of ChunDoSunBup Qi-training on growth hormone, insulin-like growth factor-I, and testosterone in young and elderly subjects. : Am J Chin Med 1999; 27(2):167-75.

Li W, Xing Z, Pi D, Li X. Influence of qi-gong on plasma TXB2 and 6-keto-PGF1 alpha in two TCM types of essential hypertension. Hunan Yi Ke Da Xue Xue Bao 1997; 22(6):497-9.

Wu WH, Bandilla E, Ciccone DS, Yang J, Cheng SC, Carner N, Wu Y, Shen R. Effects of qigong on late-stage complex regional pain syndrome. Altern Ther Health Med 1999 Jan; 5(1):45-54.

Xu SH. Psychophysiological reactions associated with qigong therapy. Chin Med J (Engl) 1994 Mar; 107(3):230-3.

Lei XF, Bi AH, Zhang ZX, Cheng ZY. The antitumor effects of qigong-emitted external qi and its influence on the immunologic functions of tumor-bearing mice. J Tongji Med Univ 1991; 11(4):253-6.

Zhou MR, Lian MR. Observation of qi-gong treatment in 60 cases of pregnancy-induced hypertension. Zhong Xi Yi Jie He Za Zhi 1989 Jan; 9(1):16-8, 4-5.

Zhang JZ, Zhao J, He QN. EEG findings during special psychical state (qi gong state) by means of compressed spectral array and topographic mapping. Comput Biol Med 1988; 18(6):455-63.

Litscher G, Wenzel G, Niederwieser G, Schwarz G. Effects of qigong on brain function. Neurol Res 2001 Jul; 23(5):501-5.

Lu YC. Biological effect of qigong waiqi − a preliminary report of the anti-injurious effect of waiqi on ozone toxicity. Zhong Xi Yi Jie He Za Zhi 1989 Dec; 9(12):734-6, 710.

Travis F, Wallace RK. Autonomic and EEG patterns during eyes-closed rest and transcendental meditation (TM) practice: the basis for a neural model of TM practice. Conscious Cogn 1999 Sep; 8(3):302-18.

Barnes VA, Treiber FA, Turner JR, Davis H, Strong WB. Acute effects of transcendental meditation on hemodynamic functioning in middle-aged adults. Psychosom Med 1999 Jul-Aug; 61(4):525-31.

Walton KG, Pugh ND, Gelderloos P, Macrae P. Stress reduction and preventing hypertension: preliminary support for a psychoneuroendocrine mechanism. J Altern Complement Med 1995 Fall; 1(3):263-83.

Jevning R, Anand R, Biedebach M, Fernando G. Effects on regional cerebral blood flow of transcendental meditation. Physiol Behav 1996 Mar; 59(3):399-402.

Elias AN, Wilson AF. Serum hormonal concentrations following transcendental meditation − potential role of gamma aminobutyric acid. Med Hypotheses 1995 Apr; 44(4):287-91.

Hebert R, Lehmann D. Theta bursts: an EEG pattern in normal subjects practicing the transcendental meditation technique. Electroencephalogr Clin Neurophysiol 1977 Mar; 42(3):397-405.

Jevning R, Wilson AF, Pirkle H, O'Halloran JP, Walsh RN. Metabolic control in a state of decreased activation: modulation of red cell metabolism. Am J Physiol 1983 Nov; 245(5 Pt 1):C457-61.

Lee MS, Huh HJ, Jang HS, Han CS, Ryu H, Chung HT. Effects of emitted qi on in vitro natural killer cell cytotoxic activity. Am J Chin Med 2001; 29(1):17-22.

D. Auricular acupuncture

Kim EH, Kim Y, Jang MH, Lim BV, Kim YJ, Chung JH, Kim CJ. Auricular acupuncture decreases neuropeptide Y expression in the hypothalamus of food-deprived Sprague-Dawley rats. Neurosci Lett 2001 Jul 13; 307(2):113-6.

Wang SM, Kain ZN. Auricular acupuncture: a potential treatment for anxiety. Anesth Analg 2001 Feb; 92(2):548-53.

Taguchi A, Sharma N, Ali SZ, Dave B, Sessler DI, Kurz A. The effect of auricular acupuncture on anaesthesia with desflurane. Anaesthesia 2002 Dec; 57(12):1159-63.

Bier ID, Wilson J, Studt P, Shakleton M. Auricular acupuncture, education, and smoking cessation: a randomized, sham-controlled trial. Am J Public Health 2002 Oct; 92(10):1642-7.

Greif R, Laciny S, Mokhtarani M, Doufas AG, Bakhshandeh M, Dorfer L, Sessler DI. Transcutaneous electrical stimulation of an auricular acupuncture point decreases anesthetic requirement. Anesthesiology 2002 Feb; 96(2):306-12.

Kim EH, Chung JH, Kim CJ. Auricular acupuncture increases cell proliferation in the dentate gyrus of Sprague-Dawley rats. Acupunct Electrother Res 2001; 26(3):187-94.

Wang SM, Peloquin C, Kain ZN. The use of auricular acupuncture to reduce preoperative anxiety. Anesth Analg 2001 Nov; 93(5):1178-80.

Tanaka O, Mukaino Y. The effect of auricular acupuncture on olfactory acuity. Am J Chin Med 1999; 27(1):19-24.

Richards D, Marley J. Stimulation of auricular acupuncture points in weight loss. Aust Fam Physician 1998 Jul; 27 Suppl 2:S73-7.

Borozan S, Petkovic G. Ear acupuncture has a hypotonic effect on the gastrointestinal tract. Vojnosanit Pregl 1996 Jan-Feb; 53(1):31-3

Shiraishi T, Onoe M, Kojima T, Sameshima Y, Kageyama T. Effects of auricular stimulation on feeding-related hypothalamic neuronal activity in normal and obese rats. Brain Res Bull 1995; 36(2):141-8.

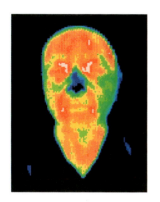 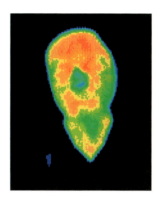

Picture 2.3.1. Acupuncture at ST36

Left: this man had hot regions around the eye and mouth, with maximum temperatures at 36.2^0 C, and 36.02^0 C (in white) respectively. Used one needle on the acupoint ST36 Zusanli.

Right: in 14 minutes, the white spots around the eyes and month had disappeared. Quantitatively, the maximum temperatures dropped to 35.76^0 C and 35.4^0 C respectively, a decrease of 0.44^0 C and 0.62^0 C. Patient reported feeling better.

Picture 3.2.1

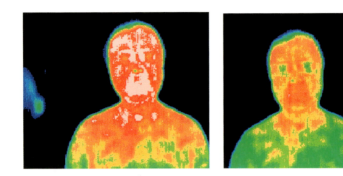

Diabetic patient with eye dysfunction

Top left: Before acupuncture treatment.
Top right: After 30 min treatment of acupuncture, many needles around the two eyes, there is great improvement.

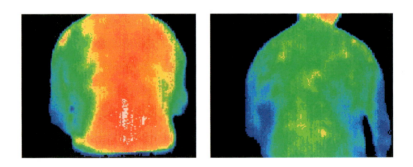

Bottom left: The back before acupuncture treatment.
Bottom right: After a 30 minute treatment the hot spots vanished.

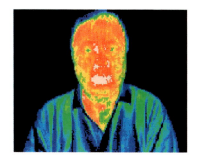 

Picture 3.2.1a: Diabetes – second treatment

Left,: before the second treatment; right: after second treatment of acupuncture without massage.

The maximum temperatures around the eyes, forehead and mouth were reduced by 0.65C, 0.26C and 0.46C after the second treatment.

The maximum temperature around the eyes, forehead and mouth were reduced by 1.32C, 1.26C and 1.0C after the first treatment. The reduction in maximum temperatures is less after the second treatment than after the first.

Before the first treatment, the differences between maximum and minimum temperatures around the eyes, mouth and forehead were 2.65C, -0.07C and 1.49C.

Before the second treatment, the differences between maximum and minimum temperatures around the eyes, mouth and forehead were 1.81C, 0.06C and 0.27.

We detected improvement around the eyes and on the forehead from the reduction of their difference in maximum and minimum temperature, from 2.65C to 1.81C and 1.49C to 0.27C. We attribute the improvement of 0.64C around the eyes, and 1.22C on the forehead, as being due to the first treatment with acupuncture and massage.

There is hardly any change in the difference of maximum and minimum temperatures around the mouth. The first treatment did not aim at any improvement around the mouth, and there is none detected. The second treatment occurred about one week after the first treatment.

The high temperature around the forehead is probably due to high blood pressure. Since the improvement is rapid, one expects that the high blood pressure can be controlled much more easily.

The high temperature around the eyes is due to damage in the eyes from diabetic illness and eye surgery operations. There is an improvement of 0.64C. This is a very good indication that the patient has responded positively to acupuncture treatment.

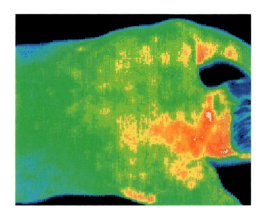

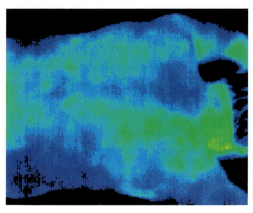

P3.2.2

Patient complained about back pain. Acupuncture given at BL40 (back knee region).

Top: before treatment; bottom: after treatment, 27 min. Maximum temperature for left middle neck reduced from 40.77C to 37.65, a reduction of -3.12C. It is also clear from the above images that the temperature of the whole back was reduced.

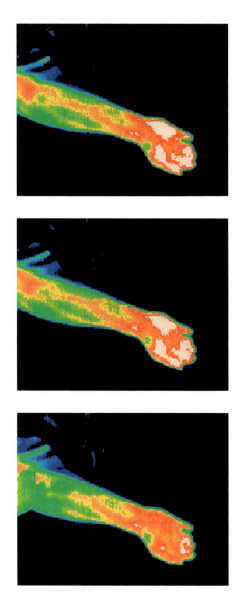

Picture 3.2.3.

Acupuncture at wrong acupoint has no effect.

Top: Before acupuncture. Middle: After acupuncture at not relevant acupoint LI13 for 1 min. Bottom: After acupuncture at relevant acupoint LU5 for 1 min.

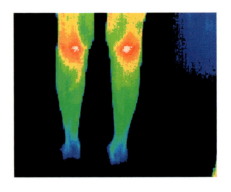

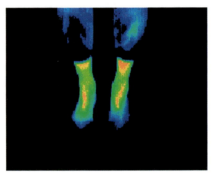

Picture 3.2.4

First picture, taken 7/14/02, reveals very cold feet. Maximum temperature on the left ankle area is 32.47C, and on the right ankle is 31.12C.

The second picture was taken on 8/4/02. After treatment on six special points of Tung's acupuncture for one week, twice a day for six out of seven days, the two legs became warmer. The maximum temperatures around the left and right ankle areas are 35.18C and 34.82C, an increase of 2.71C and 3.70C on the left and the right ankle area respectively. The maximum temperature of her head remained at 40C, with very little change. The difference between head and feet was reduced. After this treatment, the patient reported she could sleep better. The acupuncture may be the reason why. This case also illustrates the bi-directionality of acupuncture: besides cooling hot spots, acupuncture can also warm cold areas.

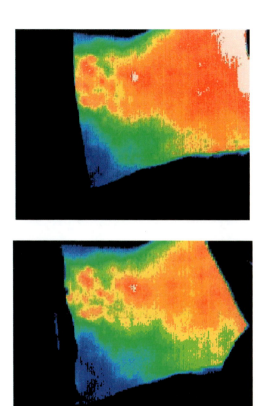

Picture 3.2.5

This is a woman patient. Top picture: before treatment. The hottest spot is at the maximum temperature of 36.59 C. Moxibustion treatment was given at DU3, DU4, BL18, BL23, BL25. Bottom picture was taken after 8 minutes of treatment. The hottest spot is reduced to 36.23C, a reduction of 0.36C.

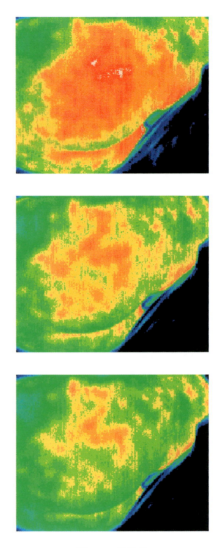

Picture 3.2.6.

Pain in the buttock: Acupuncture at GB34 (lateral part of right knee region) to treat right buttock pain.

Top: before treatment; middle: 8 minutes later; bottom: 20 minutes later. Maximum temperature reduces from 36.26C, to 35.6C to 35.25C.

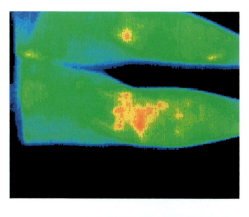
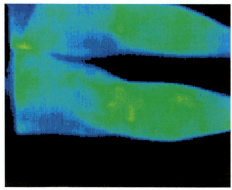

Picture 3.2.7

Pain in the feet: Patient complained of left back foot pain; treated at BL57.

Top: before treatment at BL57; bottom: after treatment at BL57 for 14 min. Maximum temperature of the right foot dropped from 36.46C to 34.75C, by -1.71C. Maximum temperature of the left foot dropped from 37.49C to 35.6C, by -1.89C.

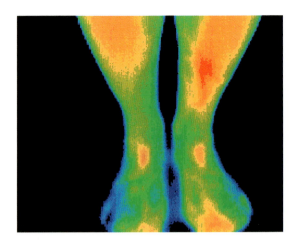

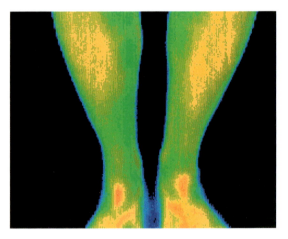

Picture 3.3.1

Balancing left and right by acupuncture using one needle

Top picture: before treatment. There is asymmetry between left and right foot. There is pain in the right foot, which shows up, in top picture, in temperature distribution.

Bottom picture shows symmetric effect of acupuncture: ten minutes after treatment with acupuncture at LR 2, which was near the toe, the high temperature area on the left foot decreased, and the low temperature of the right foot increased. The result is symmetry in temperature between left and right feet.

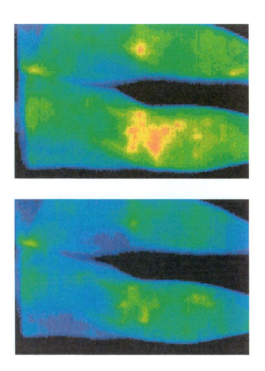

Picture 3.3.2

Balancing left and right by symmetric treatment.

Patient complained of back side left leg pain. Acupuncture was given at BL40 on both the left and right sides. Top picture: before treatment; bottom picture: after treatment of 24 minutes.

Maximum temperature of the right back knee was reduced from 37.46C to 34.68C, by -2.78C. Maximum temperature of the left back knee region was reduced from 37.59C to 35.68C, an amount of -1.91C.

This was the fifth treatment of the patient for her leg problem of twenty year's duration. The improvement was cumulative and permanent. About one treatment she said it was like magic. By the time of the fifth treatment her main complaint seemed to be gone.

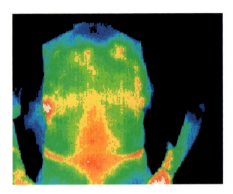

Picture 3.4.1.
Discovery of new sources of problem

Before treatment (top picture): New sources of the patient's problem were discovered. This woman, in her forties, complained about lower back pain. The thermal image revealed three problem areas. The posterior neck area was cold, the left middle back, and spinal cord area on the lower back have hot areas. Patient confirmed she had a long-standing problem in her neck, which came and went. She'd had a car accident two years ago, which caused a lump on the left middle back area. Acupuncture was administered. Moxibustion was applied in the neck region.

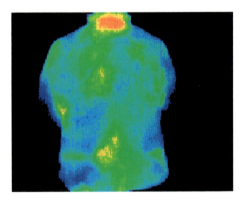

After treatment (bottom picture): Symptoms were considerably reduced in three problem areas. Repeated treatments were recommended to cure them all.

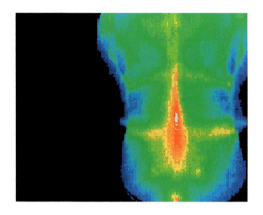

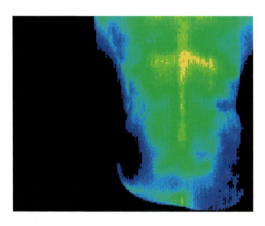

Picture 3.4.2

Exact location of pain

The top picture was taken before treatment. The woman patient had had pain for over ten years. Her son was a kidney specialist, and advised her not to take painkillers. As a last resort she came to us. We found that the origin of her pain was around the spinal cord, although her complaint was about the right leg. The doctor treated her with acupuncture at 2 Hz. After a session of twenty minutes, the spinal area showed great improvement as shown in second picture. The maximum temperature dropped from 39.4C to 37.17C, a decrease of 2.23C. See also the following pictures, P3.4.2b, describing the same patient's neck problem.

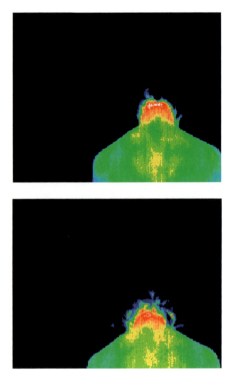

Picture 3.4.2b. Exact location of pain (cont.)

In the same patient, the neck was found to be a problem area, as indicated by the wh region in top picture, taken before treatment.

No treatment was directed toward the neck region, and there is only a little improveme probably due to the connection through the Du meridian, as shown in second picture.

The maximum temperature dropped from 38.46C to 38.17C, a decrease of 0.29C, one eigh of what it was at the lower back.

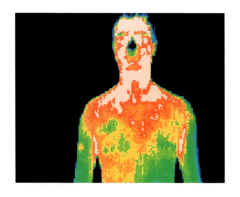

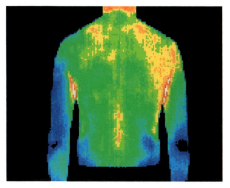

Picture 3.4.3

Example of misdiagnosis

Patient complained about lower back pain. During the prior sixteen months, he had gone to all kinds of doctors, including a chiropractor, without relief. Finally he contacted a neurosurgeon, who advised him that an operation could be performed, but that the chance of improvement was very small.

From our full scan, it was revealed that his headache problem was much more serious than the back pain. He'd had headaches for the past twenty months. Since he'd emphasized the wrong problem as being the more serious, all his physicians were misdirected and misdiagnosed him. A full body scan helped locate the most serious problem area.

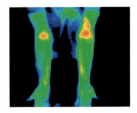 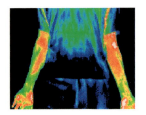

Pictures 3.4.4. Possible heart problems

Top left, person with no known heart problem. Maximum temperatures are 36.85 C and 37.49 C for the anterior right and left arms near PC3 region. The minimum temperatures are 34.54 C and 33.33 C for the anterior right and left arms respectively. The average temperature difference between the maximum and minimum temperature is 3.23 C. This figure and all subsequent figures are on the same color code that runs from 30 C – 38 C, with coldest color in black, changing to blue, green, yellow, red and to the hottest in white.

Top right: Person with no known heart problem. The maximum temperatures are 37.62 C and 37.75 C for right and left anterior arms near the PC3 regions. The minimum temperatures are 34.47 C and 35.43 C for the right and left anterior arms respectively. The average temperature difference between maximum temperatures and minimum temperature is 2.86 C.

 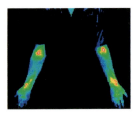

Middle left: patient with known heart problem. The maximum temperatures are 38.67 C and 38.34 C for the right and left anterior arms near the PC 3 regions. The minimum temperatures are 34.22 C and 34.44 C for the right and left anterior arms. The average temperature difference between maximum and minimum temperature is 4.18 C.

Middle right: patient with known heart problem. The maximum temperatures are 37.53 C and 37.92 C for the right and left anterior arms near PC 3 regions. The minimum temperatures are 31.85 C and 31.35 C for the right and left anterior arms. The average temperature difference between maximum temperatures and minimum temperatures is 6.13 C.

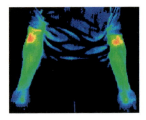

Bottom: patient after open heart surgery and fully recovered. The maximum temperatures are 37.52 C and 36.49 C for the right and left anterior arms near PC3 regions. The minimum temperatures for the right and left anterior arms are 33.08 C and 33.08 C respectively. The average temperature difference between the maximum and minimum temperature is 3.43 C.

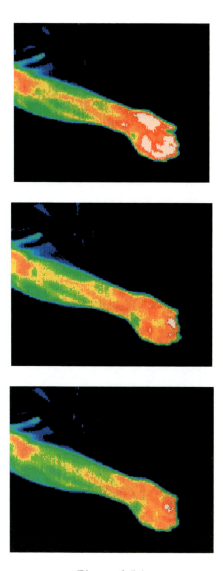

Picture 3.5.1

Temperature reduces exponentially over time: Patient suffered hand pain from repeated motion. Acupuncture given at LU5.

Top: before; middle: after 8 minutes; bottom: after 19 minutes. Maximum temperature in the palm decreased from 37.49C to 37.2C, and to 36.68C, by -1.19C. Highest temperature area (in white color) was reduced considerably.

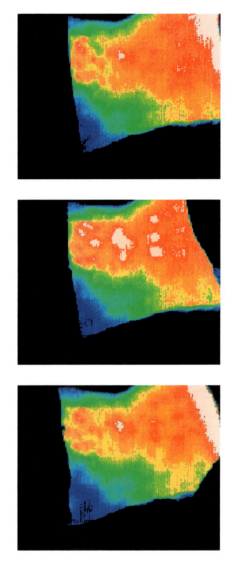

Picture 3.5.2

Moxibustion can also lower hot spots in painful areas.

Top picture: before treatment. Middle: immediately after moxibustion at BL23, 25, 18, DU3 and DU4. The back warms up as shown. Bottom picture: 2 minutes after treatment, heat due to the warming effect of moxibustion has gone and the back starts to cool off.

Table 1.2.1. Meridians as Optical Fiber - Optical Properties of Meridians

Subjects	Objective	Method	Results	Authors
muscle system of lower limb of a human body	Transparency to infrared along the meridians.	Infrared spectrophoto-meter FTS-185, American Bio-Red Co. and 5DX type, American Nicolet Co.	The axis direction of the collagenous fibre at the GB meridians has a high transparency 76% at wave length of 2.66 microns. Along the fibre axis of the specimen corresponding to ST meridian, the transparency is 62% at wave length of 9-20 microns; the transparency vertical to the axis is 0.4%.	Fei et al, Shanghai, 1998
6 normal human subjects	Thermographic visualization of changes in peripheral perfusion.	Manual acupuncture at PC 6 and PC3; thermographic recordings, Flir system, were used to assess superficial changes in temperature in the hands.	In all subjects a significant (p= 0.015) short term cooling effect during acupuncture. For long term effect, three subjects had a warming effect (> 2 degrees C), three subjects had temperature decrease.	Litscher et al, Graz, 1999
human body	Infrared thermogram on meridians.	Observation of skin temperature of central line of the back, chest and abdomen on the human body, Du and Ren meridians.	(a) 57.1% volunteers showed central lines of the back, and 7.7% showed central lines of the chest-abdomen. (b) After moxibustion at DU4, 70.4% temp of central lines of the back went up. After moxibustion at RN12, temperature of 56.0% central lines of the chest-abdomen went up; lines at the back were about 50 cm, while the lines on the chest-abdomen were about 10-30 cm.	Zhang et al, Beijing, 1996
139 healthy cases & 305 patients with facial paralysis	Imaging of facial temperature along meridians.	AGA-782 infrared thermovision and TC-800 computer for analysis.	1) Facial thermal lines appeared in 95 of 444 cases; high temperature along GB meridian in 34 cases. 2) In 26 of 95 cases, thermalline appeared after acupuncture. 3) Range of temperature was within 1.3 degrees C, in some cases only 0.2-0.3C, with +/-0.5-1.0C difference with surrounding temp; width of line was 1/1.5 cm. 4) Lines appeared stable, lasting days, weeks or a month. 5) Certain relationship exists between appearance of lines with facial paralysis.	Zhang et al, Beijing, 1992

Table 1.2.1. Meridians as Optical Fiber - Optical Properties of Meridians (cont.)

Subjects	Objective	Method	Results	Authors
23 cases	Propagated sensation along (PSC) by infrared thermal images.	Needle at Hega LI4, PC7 and PC6	10 out of 23 cases higher and middle temperature bands of upper extremity distributed along LI meridian temperature when subjects felt PSC induced by needling at LI4. None shown when they did not feel PSC by needling at other points.	Liu et al, Beijing, 1990
158 healthy subjects; 158 channels, 11,582 points	Natural luminescence phenomena along meridians and acupoints.	Ultra feeble cold luminescence signals of the body surfaces are detected.	The intensity of luminescence at the classical acupoints was 1.5 times higher than that of the matched points bilaterally 0.5 cm apart from either side of each classic channel.	Yan et al, Beijing, 1989
rats	Luminescence of meridians in rats: healthy and sick.	Detection of bio-luminescence along Ren and Du meridians of healthy rats and rats induced by hydrocortisone and by bleeding.	For heathy rats there is high emission of light along Ren and Du meridians. Rats suffering from hydrocortisone or from bleeding have low luminance along Du or Ren correspondingly. The intensity of emitted light increased after acupuncture treatment.	Yan et al, Beijing, 1992
human	Temperature oscillation spectrum	Temperature oscillation spectra in the regions of active biological points when the state of meridian is changed.	When the meridian is actviated, the enhancement of the harmonics with the period 3-4 seconds has been observed in the spectra.	Osipov et al, Russia, 1996
human	Propagation of light along meridian	White source light is illuminated on the PC6, and propagated light is detected at three points along the PC meridian and four reference points away	Light propagates better along the meridian than the reference path in more than twenty five percent for all subjects tested.	Choi et al, Seoul, preprint

Table 1.2.2. Meridians as Conductors - Evidence for Low Impedance Along the Meridians

Subjects	Objective	Method	Results	Authors
12 cases	Influence of aucpuncture on impedance.	Acupuncture at PC3 ; impedance on pericardium meridian measured by four electrodes impedance instrument.	Lower impedance along meridians from 52.8+/-11.0 ohm by 9.2+/-5.6 ohm($p<0.001$);control from 61.7+/-10.3 ohm by 0.12 =+/-2.4 ohm ($p>0.05$); implies that acupuncture makes interstitial fluid increase by axon reflection and blood capillary expansion.	Zhang et al, Beijing, 1999
humans	Distribution of low skin impedance pt (LSIP) along meridians.	Observation of LSIP along three yang meridians and ren and du meridians.	Overwhelming majority of LSIP located right on or within 5 mm bilaterally to their courses.	Huang et al, Fuzhou, 1993
12 volunteers	Distribution of low skin impedance pt (LSIP) over the medial side of forearm.	Distribution of LSIP was plotted with a comuter system designed for the measurement of skin impedance, covering the whole area of the medial side of forearm.	Most of 391 LSIP are located right on or within 5 mm bilaterally to the LI, PC, and HT meridians of the hand.	Hu et al, Fuzhou, 1993

Table 1.2.2. Evidence for Low Impedance Along the Meridians (cont.)

Subjects	Objective	Method	Results	Authors
60 healthy volunteers	Distribution of LSIP along meridians II	Studies of LSIP with a computer system. The method was reliable and repeatable.	LSIP basically distributed along the meridians.	Hu et al, Fuzhou, 1993
12 healthy volunteers	Effect of increase and decrease of measurement voltage on skin impedance.	On the medial side of forearm two levels were selected for measurement.	Impedance of LSIP decreased with increase of voltage, never exceeded 100k ohm; for non-LSIP, remained higher than 600k ohm; voltage runs between 10 to 50 v.	Wu et al, 1993
12 sheep, 13 pigs, 11 cats, 10 goats, 8 rabbits, 7 donkeys	Biophysical characteristics	Low impedance lines, high percussion sound lines.	Along the vertical planes of the dorsal line.	Yu et al, Beijing
soybean	Evidence of plant meridian system.	Connect bioelectrical potential and resistance of soybean roots, stems, leaves and pods.	higher potential and lower electrical resistance associated with vein, etc., as compared with other parts of plants; when two needles inserted in low resistance points, 26.0% lowering of electrical resistance was observed on the vein.	Hou et al, Xianjiang, 1994

Table 1.2.3. Different Effects from Different Frequencies of Electroacupuncture

Subject	Objective	Acupoints	Method	Results	Authors
rats, spinal perfusate	Frequency dependance.	EA 2, 15, 100Hz	Change of immunoreactivity (ir) of somatostatin(SOM) and calcitonin gene-related peptide(CGRP)	2 Hz increases SOM-ir 39% decreases CGRP-ir 47% 15Hz decreases SOM-ir 37%; increases CGRP-ir 92% 100 Hz inhibits SOM-ir; no effect on CGRP	Tian et al, Beijing, 1998
rat	Analgesic effect of different frequencies.	EA:2, 15, 100Hz	Nociceptive response from biting and licking.	2Hz is good for analgesia; nop effect from 100 Hz	Hsieh et al, Taichung, 2000
rats	Analgesia	2, 100Hz	Indication of nociception by radiant heat tail flick latency.	At spinal level endomorphin-I is involved for 2 Hz, but not 100 Hz in analgesia.	Han et el, Beijing, 1999
16 students	Pulse rate and skin temperature	2, 100Hz; ST36	Pulse rates were measured on the middle finger and skin temperature taken between the thumb and index fingers.	Both decreased pulse rate and temperature; 2 Hz had a more sustained effect.	Hsieh CL et al, Taichung, 1999

Table 1.2.3. Different Effects from Different Frequencies of Electroacupuncture (cont.)

Subject	Objective	Acupoints	Method	Results	Authors
rat brain	EA acelerates the expression of gene.	2 Hz, 100Hz	Release of opioid peptides in CNS.	Both increases of PPE mRNA, with different acceleratory effects with different acceleratory effects on release and synthesis of different opioids.	Guo et al, Beijing, 1997
rat	EA activates sympa-thetic nervous system.	2, 20 Hz; LI 4	Rhythmic micturition contraction, urine excretion, blood pressure, renal sympathetic nerve activity, pelvic parasympathetic nerve activity.	Selectively activate the parasympathetic nervous system; different frequencies elicit distinct mechanisms to activate sympathetic nervous system.	Lin et al, Taipei, 1998
rat	Different frequency of discharges of neurons in RVM.	2, 10 Hz; Zusanli	Unit discharge of rostral ventromedial medulla RVM neurons extracellularly by microelectrode.	Both activate excitatory neurons in RVM, suppress nociceptive transmission via opioid mechanism. Differences are observable.	Ao et al, Wuhan

Table 1.2.3. Different Effects from Different Frequencies of Acupuncture (cont.)

Subject	Objective	Acupoints	Method	Results	Authors
rat brain	Brain substance activated by EA.	2, 100 Hz	Fos-like immunoreactivity and in situ hybridization of three opioid mRNAs were used.	Distinct neutronal pathways underlying EA of different frequencies; exert differential effects on opioid gene expression; 2 Hz induces intensive preporenkephalin (PPE); 100Hz increses preprondynorphin (PPD).	Guo et al, Beijing, 1996
rat spinal cord	Frequency dependency of substance P release.	2, 4, 8, 15, 30, 100Hz	Tail flick latency; immunoreactive SP was measured by radio-immunoassay.	Decrease during 2Hz; no change for 4Hz; marked increase during 8, 15, 20 and 100 Hz.	Shen et al, Neijing, 1996
rats	Distribution of brain stem and spinal cord nuclei.	4,100 Hz on zusanli ST 36	Number and distribution of fos-immunoreactive neurons in the brain stem and spinal cord.	Both exhibited greater number of fos-labeled neurons in dorsal horn; additional brain-stem neurons are selectively activated by 4 Hz.	Lee et al, Minnesota, 1993
rats	Morphine abstinence syndrome	2, 100 Hz	Multiple pellet implantation, wet shakes, teeth chattering, weight loss, escape attempts.	100 Hz accelerates release of dynorphins in CNS; much less with 2 Hz.	Han et al, Beijing, 1993
rats	Analgesic effect	10, 100 Hz with 3 V and 6 V.	Transient analgesic after effect.	For transient, 10Hz> 200 Hz; 6 V >3V; after effect 00Hz>10Hz, 3V>6V.	Wang et al, Beijing, 1993

Table 1.2.3. Different Effects from Different Frequencies of Acupuncture (cont.)

Subject	Objective	Acupoints	Method	Results	Authors
rats	On nociceptive response	15, 100 Hz at Yanglingquan.	Analgesia spinal glutamic acid.	15Hz>10Hz for analgesia; 15Hz>100 Hz for spinal glutamic acid.	Cao et al, Beijing, 1993
rats	Different frequencies mediated by different types of opioid.	2Hz, 100Hz, 2-15 Hz.	Tail-flick latency on (?) repeated injections of ohmefentanyl OMF, or delta-opioid agonist.	2Hz mediated by mu-and delta-receptor, 100 Hz by kappa-receptor 2-15Hz by combination of the above.	Chen et al, Beijing, 1987
rats	Role of periaqueductal gray in analgesia.	2Hz, 100 Hz.	Comparision of analgesia between EA and ARH.	Hypothalamus(ARH) plays role in mediating low- but not high-frequency EA analgesia.	Wang et al, Beijing, 1990
rats	Tail flick reflex suppression.	2 Hz, 100 Hz.	Ablation of the whole forebrain (telencephalon and diencephalon).	Diencephalon for mediating 2Hz, but not 100 Hz	Wang et al, Beijing, 1990

Table 1.2.3a. Summary of Differences Between 2 Hz and 100 Hz in Electroacupuncture

Subject	Effect	2 Hz	100 Hz	Authors
rats; spinal perfusate	Immunoreactivity	Increases somatostantin (SOM) 39%, decreases calcitonin, gene-related peptide CGRP.	Inhibit SOM, and no effect on CGRP.	Tien et al, Beijing Medical University
rats	Analgesic effect; nociceptive response from biting and licking.	Good for analgesia.	No effect.	Hsieh, Taichung China Medial College
rats	Morphine abstinence syndrome, wet shakes, escape attempts, etc.	Mild suppression of syndrome, wet shakes -31%.	Increased release of dynorphins in CNS, wet shakes -61%.	Han et al, 1994
rats	Different types of opioid.	Mediated by mu and delta-receptor.	Mediated by kappa-receptor.	Chen et al, Beijing Medical University

Table 1.2.4. Meridians as Plastic Tubes with Water - Propagation of Sensation Along Meridians (PSM)

Subjects	Objective	Methods	Results	Authors
536 young-sters	Differences in response from age differences.	Comparative study for ten years.	Close relation between the degree of distinctness of PSM and the age of subjects. The incidence and distinctness of PSM was by far higher than in adults.	Yang et al, Fujian, 1993
536 young-sters	Improvement of 992 with myopic eyes.	LI4, LR3 and SJf, GB 37, punctured on alternative days, 1 to 3 periods.	868 eyes (87.5%) improved; vision completely recovered in 131 eys (13.31%); diopter decreased -0.75 to -1.00 DS. in 13 eyes. The more striking the centripetal PSM, the better the therapeutic effect of acupuncture.	Li et al, Fujian, 1993
33 subjects	Influence of mechanical pressure applied on stomach meridian.	Influence of pressing meridian course and non-meridian point on electrogastrogram EGG from two abdominal points.	Acupoint ST 36: acupuncture increased the amplitude of EGG if it was low before acupuncture and decreased it if it was high. Response is bi-directional. Compressing at ST34 markedly decreased the effect of acupuncture. Pressing on both sides of ST34 mechanically did not have any effect.	Xu et al, Fujian, 1993

Table 1.2.4. (cont.) Propagation of Sensation Along Meridians (PSM)

Subjects	Objective	Methods	Results	Authors
23 subjects with PSM, 30 subjects without PSM	Influence of pressing on electro-retinogram.	Electroretinogram ERG was recorded with surface electrodes and added by an averager.	Hegu LI4 is the acupoint; The b wave of ERG evoked by flash stimulation has a positive amplitude 22 +/- 1.66 muv with a latent period about 60ms; it changed by 26.5+/-2.43% when LI4 was punctured; it remained unchanged if pressed on Shousanli LI 10 during acupuncture; it also changed the same if pressed on both sides of LI 10 point.	Wu et al, Fujian, 1993
5 subjects with PSM, 5 subjects without PSM	Plot the large intestine meridian by blocking with mechanical pressure.	Blocking acupuncture effect by mechanical pressure using ERG as a response indicator.	Hegu LI4 was punctured; pressure was applied on LI&, LI10, LI11, LI14, and LI15, then variation of ERG amplitude decreased markedly or disappeared. Variation rate of ERG amplitude was increased if pressure was applied on both sides of any above points (p <0.01).	Wu et al, Fujian, 1993
10 subjects with marked PSM, 16 subjects without PSM	Functional characteristic of cortical somatosensory area during advance of PSM.	Short latent somatosensory evoked potential from three scalp points corresponding to foot, arm and face.	(a) Without PSM: punctured at GB43. Potential components C2 (P45-N55) attenuated from feet to face by 0.85+/-0.17, 0.51+/-0.11, and 0.31+/-0.07 micro v; punctured at LI4 component C2(P25-N55), larger in area of arm than those of foot and face. (b) With PSM: C2 was nearly the same in three areas whether stimulating LI4 or GB43.	Wu et al, Fujian, 1993

Table 1.2.5. Constituents Inside Meridians are Charged - Evidence for Meridians: Ionic and Other

Subjects	Objective	Methods	Results	Authors
volunteer's lower limbs	Calcium ion concentration.	External beam proton-induced X-ray emission (PIXE).	The vertical links between areas with concentrated calcium matched perfectly with the stomach and gall-bladder meridians.	Fei et al, Shanghai, 1998
rabbit	Detection of H+ concentration.	Rabbit suffering from arrhythmia induced by aconitine.	The concentration of H+ ions increased in PC6 as well as along the heart and pericardium meridians.	Wang et al, Tianjian, 1996
19 amputated limbs of patients and 21 rats	Morphometric observation on the mast cells.	Observation of the number, distribution and characteristics of the mast cells by microscope after stained with toluidine blue dye.	The number of mast cells was more concentrated under the meridian lines than under control areas. The difference was significant.	Zhu et al, Beijing, 1990

Table 1.2.5 Evidence for Meridians: Ionic and Other (cont.)

Subjects	Objective	Methods	Results	Authors
rats	To observe the effect of EA along meridians.	1% CB-HRP solution was injected at ST 36; HRP labeled cells were counted in the spinal dorsal root ganglia.	Sensory cell of ST 36 projected to neurons L4 and L5 DRG; that of Ruzhong projected to T4, T5 and T6 DRG. When activiated by EA, their ability to uptake HRP was enhanced; HRP labelled cells increased in related DRG; the HRP labeled DRG segments spread. This study provides evidence for the mechanism's explanation of propagated sensation along the meridians.	Xu et al, Wuhan, 1996

Table 1.2.5 Evidence for Meridians: Ionic and Other (cont.)

Subjects	Objective	Methods	Results	Authors
morpho-genetic singularity theory	Ontogeny, phylogeny and physiological function of meridian.	Embryogenesis	As an embryo develops into comunication compartment domains, boundaries become major pathways of bioelectrical currents. Separatrices can be folds on the surface or boundaries between different domains. Meridians are separatrices and related to an underdifferentiated, interconnected cellular network that regulates growth and physiology, e.g., part of LU meridian runs along the borders of the biceps and branchioradialis muscles; part of the pericardian meridian runs between the palmaris longus and flexor carpi radialis. muscles.	Shang, Emory University 1999

Table 1.3.1. Properties of Acupoints - Acupoints as Water Well, or as Local Organizing Centers

Subject	Purpose	Method	Results	Authors
A. Low Electrical Resistance				
6 rats	Topography of low skin resistance points.	A method employing the unijunction transistor relaxation oscillator for detecting low skin resistance points.	The resistance on two major meridians : 14 Ren acupoints and 17 Du acupoints are in the ranges of 179.4+/-41.2 K ohm and 152.5+/-32.2 K ohm respectively. Most other oints have resistance greater than 420K ohm.	Chiou et al, Taipei, 1998
men and rats	Skin resistance and the size of acupoints.	Measurement of size of area of increased skin conductivity (ASIC), which corresponds to the localization of acupoints.	The approximate size of AISC is 350 microns for rat and 450 microns for human.	Jakoubek et al, 1982
human	Charaterisation of human skin conductance at acupoints.	Quantification of the skin conductance under well defined conditions, using the electrode materials gold, graphite, silver and brass.	The observed current response appeared to be best described by two exponentials.	Comunetti et al, Basel, 1995

Table 1.3.1. Properties of Acupoints (cont.)

Subject	Purpose	Method	Results	Authors
B. Structure of Acupoints				
experimental animals	Ultrastructures of EA or moxibustion stimulated skin and regional lymph nodes.	Examination by scanning and transmission electron microscope by moxibustion. [makes no sense]	Moxibustion-treated instances revealed a large number of immunocyte infiltrations in the region. They consisted of lymphocytes, monocytes and some granulocytes and mast cells. Lymph nodes increased in weight and induced numerous immunocyte influxes through afferent lymphatics.	Kimura et al, Osaka, 1998
157 subjects	Absorption of Tc-99 pertechnetate via acupoints and	A new method of radionuclide venography by subcutaneous injection of Tc-99m pertechnetate (SC-RNV) at acupoint Ki-3 Taixi.	Absorption of radioisotopes via KI3 was better than that via non-acupoint, evidenced by higher peak activity and greater absorption rate.	Wu et al, Kaohsiung, 1996
57 cases with	Diagnosis of deep vein thrombosis (DVT)	SC-RNV at acupoint KI3	Proved it to be superior to radionuclide venography by intra-venous injection as a lower limb venography.	Wu et al, Kaohsiung, 1993
human	Differences between SP10 point and non-acupoint.	SC-RNV; time-acitivity curve of Tc-99m.	Absorption was significantly greater than at non-acupoint, evidenced by higher peak activity and greater accelerating rate.	Chen et al, Kaohsiung, 1993

Table 1.3.1. Properties of Acupoints (cont.)

Subject	Purpose	Method	Results	Authors
human	Relation of acupoint with blood vessels.	SC-RNV at Ki 3 and BL 60	A lower limb venography was obtained similar to intravenous injection; suggested that some acupoints do play a role in drainage of tissue fluid from soft tissue into the veins.	Wu et al, Kaoshiung, 1990
11 males and 11 females	Comparison with venography by intravenous injection.	SC-RNV at KI3 and BL60 to detect varicose veins and partial stenosis of the deep veins, complete stenosis of the deep veins.	SC-RNV showed almost the same results as IV-RNV in 21 (47.7%) cases, superior to IV-RNV in 22 (50%) cases and inferior to IV-RNV in 1 (2.3%) case.	Wu et al, Kaoshiung, 1989
one detached lower limb and three whole human body specimens	Physical basis of meridians and acupoints.	Magnetic resonance image location of acupoints; anatomical locations, 73 acupoints; XCT location; Hi-Scope Video Microscope system.	It is a complex system mainly of connective tissue and is interwoven with the blood capillaries, nerves, lymph vessels, etc. Elements of Ca, P, K, Fe, Zn, Mn, etc. are found concentrated in the deep connective tissue structures in locations corresponding to acupoints.	Fei et al, Shanghai, 1998
morphogenetic singularity theory	Acupoints originate from the organizing centers in morpho-genesis.	Ontogeny and phylogeny	This theory explains the distribution and non-specific activation of organizing centers and acupoints, the high electrical conductivity of acupoints, the polarity effect of EA, and the physiological functions of meridian and chakra systems.	Chang C, Emory U, 1999

Table 1.3.1. Properties of Acupoints (cont.)

Subject	Purpose	Method	Results	Authors
rats	Influence of EA on mast cells.	EA at Biguan and zusali, and observed subcutaneous mast cells of Tianshu, Biguan, Futu, Zusali and Xiajuxu.	The number of mast cells in all the acupoints studied were higher than that in the non-acupoint areas.	Deng et al, Wuhan, 1996
rats	Effect on mast cell from EA at ST36.	After unilateral transection of sciatic nerve, the number N of subcutaneous mast cells and its degranulation in ST36 and Xiajuxu was studied.	Sciatic nerve transection could reduce N; hence suggested peripheral nerves play an important role in the convergence and degranulaton of mast cells in acupoints by EA.	Deng et al, Wuhan, 1996
human	Acupoint tissue constitution	Twisting-needle manipulation to acupoints BL23, 24, 25 for an induction of Qi; transparent materials were binding to the needle.	Electron-microscopical analysis of the transparent materials revealed that they were made up of collagen fibers, elastic fibers, fibroblasts, adipocytes and mast cells. Rarely were nerve fiber-like structures observed. Nevertheless, calcitonin gene-related peptide-positive nerve fibers could be demonstrated in the acupoint BL24-associated fascia.	Kimura et at, Osaka, 1992

Table 1.3.2. Effect of Laser Acupuncture - Photonic Stimulation of Meridians

Subject	Purpose	Method	Results	Authors
40 Children, with	Compare desmopressin with Laser Acupuncture	One group with desmopressin and another group with laser acupuncture, initially with 5.5 wet nights per week.	Desmopressin group has a complete success rate of 75%, and 20% no response; laser acupuncture group has a complete success rate of 60%, and 15% no response. Conclude laser acupuncture is as good as well-established desmopressin therapy.	Radmayr et al, Innsbruck, Austria, 2001
children	Reduces post-operative vomiting	Laser stimulation at P6.	The inidence of vomiting was significantly lower 25% than that in the placebo group 85%.	Schlager et al, Innsbruck, Austria, 1998
87 patients	With facial neuritis (correction of paralysis).	Laser with wavelength 890nm with intensity 20m@ on 10 to 15 points, 1 to 3 min, 12 to 16 sessions.	Good clinical effect confirmed by results of studies made with the aid of electrophysiological modalities.	Chupryna, Ukraine

Table 1.3.2. Effect of Laser Acupuncture (cont.)

	Purpose	Method	Results	Authors
rats	Effects on the spleen in rats.	The splenic sinuses, lymphocytes and antigen-presenting cells in the spleen of rat stimulated by He-Ne laser acupuncture were observed by using TEM.	Width of endothelial interstice was increased; a lot of blood cells oozed out of the splenic sinuses; numerous activated T cells; B cells were gradually differentiated into large lymphocytes; macrophages had short process with numerous folds and microvilli and tended to neighboring lymphocytes. Hence the activities of the cellular immunity and humoral immunity were enhanced by laser.	Dong et al, Hebi, 1996
Dutch hybrid	Analgesia effect evaluated by somato-sensory evoked potential.	A gallium arsenide laser diode with wavelength 780nm, switched pulse frequency 9720Hz, 0.6J.	Tooth pulp generated somato-sensory evoked potential recordings showed decreased peak-wave amplitude in late near-field components. These decreases correlate to analgesia.	Sing et al, Hong Kong, 1997
162 patients	Treatment of chronic cerebro-vascular disorder.	Infrared portable laser for 10-12 procedures; microclimate of biotron.	70 % of patients with initial manifestations of cerebral blood supply insufficiency derived benefit from laser acupuncture.	Macheret et al, Korkushko, 1996

Table 1.3.3. Effect of Transcutaneous Electric Nerve Stimulation at Acupoints

Subject	Purpose	Method	Results	Authors
12 patients	Effect of TENS on therapy-resistant hypertension.	Treated with TENS at two acupoints on both forearms.	After 4 weeks of TENS, the mean diastolic blood pressure decreased by 3.7 mm; the mean systolic blood pressure decreased by 6.3 mm ($p < 0.05$). The effect of TENS may also have a prolonged effect.	Jacobsson et. al, Sweden, 2000
100 women after hysterectomy or myomectomy	Effectiveness on postoperative hydro-morphone (HM) requirement.	TENS at standard dense and dispersed mode of 2/100 Hz at dermatomal at level of surgical incision, or at ST36; used patient controlled analgesia device to deliver bolus doses of HM 0.2-0.4 mg on demand.	The opioid requirement was decreased by 35% and 38%. TENS at the dermatomal level of skin incision was as effective as at STS36, and both were more effective than at nonacupoint (shoulder) location.	Chen et. al, U of Texas, Dallas, 1998

Table 1.3.3. Effect of Transcutaneous Electric Nerve Stimulation at Acupoints (cont.)

Subject	Purpose	Method	Results	Authors
22 diabetic rats	Compare EA and TENS at Shenshu and Zusali	20 min. once every 2-3 days for five weeks; diabetes induced by injection of streptozotocin, 50 mg/kg.	Increased plasma glucose levels were lowered in EA, $p<0.05$, and slightly in TENS, $p<0.05$. Symptoms of polyphagia, polydipsia and polyuria were attenuated in EA group. The motor nerve conduction velocity lowering was prevented in EA and TENS group. In TENS group pain threshold was lowered, but was elevated in EA group. Efficiency of EA treatment on experimental diabetes and its neuropathy was better than that of TENS therapy.	Mo et al, Beijing, 1996
101 women after low intra-abdominal surgery	On postoperative analgesic requirement.	Transcutaneous acupoint electric stimulation (TAES), alternating 2 and 100Hz every 3 s, placed at Hegu; patient-controlled analgesia (PCA) requirement of hydromorphone (HM).	High intensity (9-12mA) decreased HM requirement by 65% and reduced the duration of PCA therapy, as well as the incidence of nausea, dizziness and pruritis.	Wang et al, U of Texas, Dallas, 1997

Table 1.3.3. Effect of Transcutaneous Electric Nerve Stimulation at Acupoints (cont.)

Subject	Objective	Method	Results	Authors
10 healthy volunteers	Transcutaneous nerve stimulation (TNS) on esophageal function.	TNS was applied to esophageal acupoints; examined esophageal motility and vasoactive intestinal peptide levels.	Hand TNS improved lower esophageal sphincter (LES) relaxation and percent of peristaltic contractions to swallow, and decreased the number of spontaneous contractions; conclude that a somatovisceral pathway involving the esophagus exists.	Chang et al, U of Pennsylvania, 1996
human subjects	Analgesic effects induced by TENS and EA with different electrodes.	Two pairs of electrodes were placed around the point of deep pain measurement.	TENS with surface electrodes significantly increased pain thresholds of skin and fascia, but not those of muscle or periosteum; EA with non-insulated needles induced a greater increase in pain threshold in skin, fascia and muscle. EA with insulated needle was the only technique that produces a significant increase in pain threshold in muscle and periosteum.	Ishimaru et al, Kyoto, 1995

Table 1.3.3. Effect of Tanscutaneous Electric Nerve Stimulation at Acupoint (cont.)

Subject	Objective	Method	Results	Authors
32 patients with spinal spasticity	TENS as treatment.	Han's acupoint nerve stimulator (HANS) on acupoints on the hand and leg; high frequency 100 Hz, but not low frequency 2 Hz, was effective in ameliorating muscle spasticity.	Suggests anti-spastic effect elicited by peripheral electrical stimulation is mediated by the endogenous opiod ligand interacting with the kappa opiate receptors, most probably dynorphin, in the central nervous system.	Han, et al, Beijing, 1994

Table 1.3.4. Evidence of Acupressure - Mechanical Stimulation of Acupoints

Subject	Purpose	Method	Results	Authors
gagging dental patients	Antigagging effect	Thumb pressure at P6.	Not as effective as acupuncture. A substantial percentage of gagging patients would be able to go through dental procedures without gagging if the P-6 point were stimulated.	Lu et al, U of Penn, 2000
97 women	Against morning sickness, alleviating nausea and vomiting.	Use of wrist bands.	71% of women in the intervention group reported both less intensive morning sickness and shorter duration of symptoms, 0.85 hr; placebo group: 59% reporting less intensity and 63% shorter duration, 2.74 hr ($p=0.0018$).	Norheim et al, Norway, 2001
60 women	Effect on nausea and vomiting during pregnancy (NVP).	Acupressure at P6	Reduce NVP significantly with acupressure at P6 as compared to acupressure at a placebo point or none at all.	Werntoft et al, Sweden, 2001
25 healthy subjects	To relieve motion sickness and abnormal gastric activity.	Wrist band, Acuband. Subjective symptoms of motion sickness and abnormal gastric activity, as recorded via electrogastrography.	Subjects reported significantly fewer symptoms of motion sickness on days when wearing Acuband.	Stern et al, Penn State U, 2001

Table 1.3.4. Evidence of Acupressure (cont.)

Subject	Purpose	Method	Results	Authors
children	Effect on postoperative vomiting after strabismus surgery.	Applying acupressure disc on Korean hand acupuncture point.	In the acupressure group the incidence of vomiting was significantly lower (20%) than in the placebo group (68%).	Schlager et al, Innsbruck, 2000
24 healthy male volunteers	Effect on cardiovascular system	Pressure on acupoints (P) vs. stroking along the meridians.	There was a decrease in systolic arterial pressure, diastolic arterial pressure, mean arterial pressure, heart rate and skin blood flow in P group.	Felhendler et al 1999
17 women	Effect on women undergoing chemotherapy	Finger acupressure bilaterally at P6 and ST36; nausea experience measured by the Rhodes Inventory of Nausea, Vomiting, and Retching, and nausea intensity.	Significant difference found in regard to nausea experience ($p<0.01$) and nausea intensity ($p<0.04$) during the first 10 days of the chemotherapy cycle, in favor of acupressure.	Dibble et al, UC San Francisco, 2000
84 subjects	Improvement of the quality of sleep.	Acupressure therapy; Pittsburgh Sleep Quality Index; effects on hypertension, hypnosis, naps and exercise.	Significant reduction in the frequencies of nocturnal awakening and night wakeful time in the acupressure group; confirmed effectiveness of acupressure in improving the quality of sleep of elderly people.	Chen et al, Taipei, 1999

Table 1.3.5. Evidence of Injection at Acupoints - Stimulation of Acupoints with Foreign Substance

Subject	Purpose	Method	Results	Authors
91 patients with verruca	Effect of B12 injection into acupoints.	Injection of B12 into patients who had unsuccessful therapy of more than three methods.	Complete cure, 35 cases; significantly improved, 56. Total effective rate is 100%, compared with Vitamin B12 i.m. injection and polyinosinic cytidylic acid i.m. injection, cant, $p<0.001$. Also proved by histologic examination.	Hu et al, Beijing, 1998
mice	Bee venom (BV) on visceral anti-nociception.	BV injection into Zhongwan RN12.	BV acupoint stimulation can produce visceral antinociception.	Kwon et al, Seoul U, 2001
rodent arthritis	Antinociceptive effects of BV injection.	High (1:100) and low (1:000) dose of BV injection into acupoint Zhongwan RN12 and non-acupoint.	Injection into RN12 was found to produce significantly greater antinociception than non-acupoint (10 mm away from RN12) in the abdominal stretch assay. Injection into ST36 produced more potent antinociception than non-acupoint; high dose BV injection produced potent antinociceptive effect irrespective of location.	Kwon et al. Seoul U, 2001
rats; adjuvant-induced arthritis	Effect of BV injection on arthritic pain.	BV injection into Zusali ST36 (1mg/kg per day).	Dramatically inhibits paw edema; has both anti-inflammatory and antinociceptive effects; promising therapy for long-term treatment of rheumatoid arthritis.	Kwon et al, Seoul U, 2001

Table 1.3.5. Evidence of Injection at Acupoints (cont.)

Subject	Purpose	Method	Results	Authors
56 cases of biliary colic	Role of Qimen in treatment.	Water injection in Qimen LI14, BG24, RN14.	Pain disappeared in 32 (57.1%) cases treated. Total effect rate was 96.4%, better than the control group treated with Western medicine, $p<0.05$.	Jiang et al, Sichuan, 1995
54 cases of rheumatoid arthritis (RA)	Effect on immunological function.	Warm needling and point-injection with Zhuifengsu.	Effective rate was 100%; Nk activity and IL-2 value in RA patients were found to be lower than those of normal individuals; both increased after treatment. This treatment exerted a regulatory effect on the cellular immunological function.	Liu et al, Beijing, 1993
36 cases of senile dementia	Combined treatment of acupuncture and acupoint injection.	Acupoint injection of ace-glutamidi with acupuncture.	Effective for multi-infarct dementia; the rate of success being 42.85% and of improvement 42.86%; total efficacy rate being 85.71%. The rating was based on revised Hasegawa Dementia Scale and the Functional Activity Questionnaire. High density lipid cholesterone increased significantly after treatment.	Chen Y, 1992
102 cases of allergic rhinitis	Desensitising teatment.	Using acupuncture on endermic points of the head with the extract of positive allergens.	Diameter of redness and swelling on the skin was significantly reduced, $p<0.01$; cell mediated and humoral immunity of the patient tended to be normal. 72.8% of the cases had significantly curative effect; 23.56% turned better, $p<0.01$.	Zhou et al, 1991

Table 1.3.6. Evidence of Moxibustion - Meridians as Infra-red Transparent Optical Fiber

Subject	Purpose	Method	Results	Authors
A. On Immune System				
rats; ulcerative colitis. total number=32	Expression of IL-1[beta] and IL-6 mRNA.	The SD rat ulcerative colitis model was created by immunological method associated with local stimulation. Moxicones of mugwort floss were placed on medicinal pad for Qihai RN6 and Tianshu ST25 and ignited.	IL-1[beta] and IL-6 mRNA were markedly lower in the moxibustion group than in the model group; hence moxibustion inhibited the expression IL1 (beta) and IL-6 mRNA in experimental ulcerative colitis rats.	Wu et al, Shanghai, 1999
sarcoma: S180 ascitic mice	Effect on erythrocytic immunity.	Moxa-cone moxibustion at Guanyuan RN4.	Increase the decreased erythrocytic C3b receptor rosette forming rate; lower the raised erythrocytic immunocomplex rosette forming rate, increase activity of erythrocytic immunosuppressive factor in the tumor-bearing mice. Hence, moxibustion strengthens erythrocytic immunity and promote its regulative function.	Wu et al, Chenga U, 2001
tumor bearing mice	Cellular immune function.	Moxibustion at Guanyuan RN4.	Instant elevation of serum ACTH and beta-EP caused by moxibustion. Regulation of beta-EP and ACTH was related to the immune regulation induced by moxibustion.	Zhai et al, Shanghai, 1996

Table 1.3.6. Evidence of Moxibustion (cont.)

Subject	Purpose	Method	Results	Authors
tumor bearing rats	Regulation on beta-END	Moxibustion at RN4.	Moxibustion promotes the hyperplasia of the pituitary and adrenal gland, which showed atrophy in control group; also stimulated the secretion of beta-END from the pituitary and adrenal gland, increased the level of serum beta-END significantly and kept the high level for quite a long time.	Zhai et al, 1994
adjuvant arthritis rats	Anti-inflammatory and immune regulation.	Moxibustion at BL23 Shensu.	Moxibustion can lighten a local inflammatory reaction, eliminate swelling, prevent or reduce the polyarthritises, maintain the weight and shorten the course of the disease. It can recover and promote the effects of Concanavalin-inducing splenic lymphocyte proliferation in rats. It can also promote interleukin2 production, and decrease IL-1 contents.	Tang Z, Hefei, 1996

Table 1.3.6. Evidence of Moxibustion (cont.)

Subject	Purpose	Method	Results	Authors
B. On Analgesia				
Wistar rats, urethane-anesthetized	Noxious inhibitory controls (DNIC) in analgesic mechanism.	Moxibustion induced-analgesia. Single unit extracellular recordings from neurons in the trigeminal nucleus caudalis were obtained from a micropippete; 52 single cells were examined.	Suppression was observed on both wide dynamic range (WDR) and nociceptive specific (NS), but not on low-threshold mechanoreceptive (LTM) units. Moxibustion induced moderate suppression with a long induction time. Suggests DNIC may be involved in analgesic mechanism.	Murase et al, Meiji U, Kyoto, 2000
rats	Analgesic effect of moxibustion.	Radiant heat exposure on Ciliao point and measurement of latency of tail flick threshold LTH.	When temperature of surface of point was modulated within 38-39 C and 43-44C, LTH increased 17.7 +/- 2.1% and 22.2+/-2.5% after 5 min and by 16.1+/-2.9%, 22.1+/-3.4% and 21.9+/-3.2% (50-52 C) after 10 min; P<.05.	Fang et al, Beijing, 1993.

Table 1.3.6. Effects of Moxibustion (cont.)

Subject	Purpose	Method	Results	Authors
C. On renal function, colitis, ulcer, neurons and gene expression.				
Spontan-eously hyper-tensive rats	Regulation of renal function and secretion of hormone.	Moxibustion at BL15 abd BL27.	Urinary volume increased for BL15, but decreased for BL27; urinary excretion of Na+ decreased for BL15 and BL27; systolic blood pressure decreased for BL15, but not for BL27. Plasma levels of aldosterone and renin activity incrased; atrial natriuretic peptide decreased for BL15. Plasma levels of aldosterone and atrial natriuretic peptide increased for BL27.	Lee et al, Wonkwang U, Korea, 1997
mice	Effect of moxibustion on the function of MDR gene product P-glycoprotein P-170.	Moxibustion at RN 4 of BABL/c mice with S-180R adriamycin-resistant tumor cells; analysis of the drug accumulation in the S-180R cells by flow cytometer.	A weak inhibition was found when moxibustion at RN4 alone, and a very significant inhibition was observed in the presence of low-dosage of verapamil, but not at high dosage.	Zhang et al, Beijing, 1994
62 patients	Efficacy of moxibustion and acupuncture on chronic colitis.	Moxibustion at Tianshu ST25 and Guanyuan RN4.	Moxibustion had a marked curative effect with few side effects.	Yang et al, Liuzhou, 1999

Table 1.3.6. Effects of Moxibustion (cont.)

Subject	Objective	Method	Results	Authors
rats, 18	Effect on gastric mucosa.	Moxibustion on experimental gastric ulcer at BL23 Shenshu.	1. Significantly reduced the ulcer area, $p<0.05$; 2. Increase in copper content and Cu/Zn ratio; 3. Significant increase in Zinc content in serum; pretreatment by moxibustion has protective effect on gastric mucosa	Chen et al, Zhenjiang, 1995
rats	Gastrointestinal motility of reserpinized rats.	Stimulating Zusali ST36 with moxa-stick.	Strong stimulation can increase the activity of cholinesterase, $p<0.05$; inhibit hyperactive gastrointestinal motility, $p<0.05$; maintain normal body temperature and prevent body weight loss.	Liu. 1995
male rats, 33	Protection against liver injury.	Use moxibustion for rats treated with oral alphanaphthylisothiocyanate (ANIT, 100mg/kg).	Moxibustion therapy may be able to prevent ANIT-induced hyperbilirubinemia and cholangitis.	Yang et al, Taichung, 1993
rats	Effects on primary sensory neurons in the skin protecting gastric mucosa in rats.	Use immunocytochemisty combined with a fluorescent retrograde tracer dye, gluoro gold.	Moxibustion induced galanin expression by primary sensory neurons containing substance P.	Kashiba et al, Osaka, 1992
rats	Protecting gastric mucosa in rats.	Moxibustion at BL23.	Moxibustion pre-treatment within 3 weeks significantly prevented the formation of gastric ulcers.	Chen et al, Jiangsu, 1992

Table 2.3.1. Meridians in Plants - Meridians as Most Primitive & Fundamental System

Plants	Parts	Methods	Results	Authors
soybean	roots, stems, leaves, pods	Two needles into two low-resistance pt. leaf cushion area.	Electrical resistance reduces 26.0% on the main vein, 4.5% on the mesophyll of the soybean leaf for five hours.	Hou et al, I
soybean (Glycine max)	unifoliolate buds	Two needles opposite each other in the stem for whole period of experiment.	Main vein temperature increased 0.59C and 0.48C first and second days; mesophyll temperature increased 0.50C and 0.47C first and second days.	Hou et al, II
phylodendron	leaves	Laser beam on leaves	Leaves produce sound 50Hz to 120Hz; absorb external sound stimulaton below 150 Hz.	Hou et al, III
		Drought	Sound emission increased 20-30dB; response to external stimulation decreases 10-20dB; returned to normal 6 min after watering	
		Needle into petiole	Spontaneous sound production increased 40 dB in the main vein, 6dB for mesophyll.	
Phaseolus vulgaris L., pole bean and bush bean	growth and development	Two needles opposite side of the stem of unifoliolate buds.	Photosynthesis rate increased 20.5%, transpiration 27.2%; length of internodes 22.5%; dry weight of shoots from cotyledon to apex, 22.9%. Flowered three days earlier; 14.4% more fruit.	Hou et al, IV

Table 2.3.1. Meridians in Plants - Meridians as Most Primitive & Fundamental System

Plants	Parts	Methods	Results	Authors
Phaselus vulgaris L. pole bean	circumnutation movement of shoots	Two needles opposite sides of the stem near univoliolate bud for the whole experiment.	Period of ultradian rhythms of circumnutation movement reduces: (1) First group 124.2 min (control) to 116.3 min (treated) (2) Second group 132.1 min (control) to 96.7 min ($p<0.0001$)	Hou et al, V
Tomato	yield and quality	Agri-wave technology: broadcast sound wave of certain frequency and compound microelement fertilizer.	Comparison of treated and control: 1. Wt of ripe tomatoes increases 30.73% 2. Wt of unripe is 27.29% less 3. Yield is 13.89% higher 4. Storage period almost double 5. Increased sugar content by 26.19% 6. Vitamin A increased by 55.3% 7. Niacin increased by 92.31% 8. Vit. C and E decreased by 2.1% and 12.69% 9. Among 33 minerals, 26 increased in content and 7 decreased.	Hou et al

Table 2.3.2. Various Effects of Stimulating ST36 Zusanli

Subject	Purpose	Methods	Results	Authors
344 rats	Stimulation on splenic natural killer (NK) cytotoxicity	NK cytotoxicity measured by 4-h 51 Cr release assay; endogenous cytokine activities in aqueous spleen extracts. EA at ST35, 1 Hz, 3 days.	Enhanced splenic NK cytotoxicity ($p < 0.001$); high levels of interleukin (IL)-2 and interferon (IFN)-gamma ($p<0.01$); positive correlation between cytokine and splenic NK cytotoxicity, $p<0.01$. IL-2 and IFN-gamma may play a part in regulating NK cell activity.	Yu et al, Showa U, Tokyo, 1997
stressed Wistar rats	High blood pressure (BP); blood viscosity (BV).	Elevation of BP and BV by fixing and hanging EA and microinjection of Baga (60 mcgm/10 ml) EA at ST36.	Depressant effect of EA at Zusali on high BP and blood hyperviscosity may be medited by activation of GABAA receptors in the brain.	Jin et al, Shanghai, Med U, 1992
rats	Mechanism of regulation of gastrointestinal function of EA.	Use radioimmunoassay to measure immunoreactive beta endorphin (ir-beta-Ep) in gastrointestine, pituitary and plasma EA at ST36.	Contents of ir-beta-Ep elevated in gastroparietal mucous membrane, pylonic mucous membrane, duodenum, jejunum and ileum, but not in pituitary and plasma; suggests Zusali regulates the function of gastrointestinal tract through endogenous opiate peptides in gastrointestine .	Yang et al, 1989

Table 2.3.2. Various Effects of Stimulating ST36 Zusanli (cont.)

Subject	Purpose	Methods	Results	Authors
rats; 52	Discharge-inhibitory reaction induced by distending stomach.	Unit discharges of neurons in feeding center of lateral hypothalamic area (LHA) EA at ST36.	36 (69.2%) showed abolishing the inhibitory reaction induced by distending stomach; Zusali may influence the reactivity of stomach through feeding center of LHA.	Wu et al, Guilin, 1990
rats; 40 males	Effect on serum endocrine hormones by moxibustion.	Measure serum values of TSH, T3, T4, FSH, LH, testosterone, and insulin with radioimunoassay. Moxibustion at ST36.	Significant increase in insulin (p<0.01); decrease in T4(p<0.01); no marked difference in others. Hence moxibustion may affect neuro-endocrine system, improve microcirculation of pancreas gland and inhibit function of thyroid gland.	Zhang et al, 1989
rats, endotoxic shocked	Adrenal gland	EA at Ren zhong or Zusali; quantitative histochemical changes of glycogen SDH in adrenal gland; EA at ST36.	Improve the function of adrenal cortex and achieve recovery.	Wang et al, Beijing, 1996

Table 2.3.2. Effect of Stimulating ST36 Zusanli (cont.)

Subject	Purpose	Methods	Results	Authors
rats	To find the origin of nitric oxide synthase (NOS) around ST36.	Combined method of retrograde transport of horse radish peroxidase and nicotinamide, adenosine dinucleotide, and phosphate diaphorase.	Some peripheral processes of NOS positive neurons were distributed in the zusalin point from ganglia of L4 to S1, and some were projected from lamina IX of L4 to S1 in spinal cord. Conclusion: NOS positive nerve fibers in the Zusali area might be one of the morphological foundations of acupuncture effect of Zusali.	Xiong et al, Anhui, 1998
rats	Induce convergence and degranulation of sub-cutaneous mast cells.	Unilateral transection of sciatic nerve; EA at ST36.	Sciatic nerve transection could reduce the total number of mast cells and degranulation of numbrs of mast cells in Zusanli. Suggests peripheral nerves play an important role in the convergence and degranulation of mast cells in acupoints.	Deng et al, Tongji Med U, Wuhan, 1996
rats	Connection among ST36-spinal dorsal horn neurons, trigeminal sensory nucleus (TSN).	Microelectrode recording from	(a) Single spinal dorsal horn neurons receives ST 36 afferent input and then conveys it to the TSN; (b) Some spinal dorsal horn neurons receive, in turn, ennervation from TSN; (c) The convergence and integration between ST36 and TSN inputs might occur in the spinal dorsal horn neurons.	Meng Z, Beijing, 1995

Table 2.3.2. Various Effects of Stimulating ST36 Zusanli (cont.)

Subject	Objective	Methods	Results	Authors
rats; 36 males	Effect on carbon tetrachloride (CC14) induced acute liver injury.	Acupuncture at ST 36 and Tai-Chung LI3.	Higher levels of serum glutamate-oxalate-transaminase (sGOT) and serum glutamate-pyruvate-transaminase (sGPT) from CC14 induced injury were significantly reduced; acupuncture is effective in treating CC14 induced liver injury.	Liu et al, Taoyan, 2001

Table 3.1.1. Effect of Acupuncture on Immune System

Subject	Objective	Method	Results	Authors
Human brain	Immune modulation.	Maps of magnetic resonance imaging.	Reveals marked signal decreases bilaterally in multiple limbic and deep gray structures, including the nucleus accumbers, amygdala, hypothalamus, hippocampus, and ventral tegmental area; suggests immune modulation during acupuncture.	Gollub, BL et al, Massachusetts, 1999
38 patients with allergic asthma	Immunomodulatory effect of acupuncture.	Peripheral blood parameter: eosinophils, lymphocyte sub-populations, cytokines, in vitro lymphocyte proliferation.	Within lymphocyte subpopulations the CD3+cells (p=.05) and CD4+ cells (p=.104) increased significantly. Significant changes in cytokines: interleukin IL-6 (p=.026) and IL-10 (p=.001) decreased, while IL-9 (p=.050) rose significantly. In vitro lymphocyte proliferation rate increased significantly (p=.035); eosinophils decreased from 4.4% to 3.3%. Hence acupuncture showed significant immune modulating effects.	Joos et al, Heidelberg, Germany, 2000
rats	Effect of intrathecal morphine and EA on cellular immune function.	intrathecal injection of (ITH) morphine and EA on ST36 and Extra 37.	EA could prevent the decrease of lymphocyte proliferative response by ITH morphine; same tendency was observed on the induction of IL-2 production.	Sun et al, Shanghai, 2000
90 patients with painful disorders	Effect on immune response related to opioid-like peptides.	Treated at ST36 and LI4	(1) Considerable increase of beta-endorphin ; (2) significant increse of CD3, CD4, and CD8; (3) monocyte phyagocytosis was increased in 100% of treated subjects; the percentage of NK cells was increased in 50% of the treated subjects.	Petti et al, Roma, 1998

Table 3.1.2. Effect of Acupuncture on Nervous System

Subject	Objective	Method	Results	Authors
rats	Regeneration of 10mm gap of sciatic nerve.	Acupuncture and electroneedling; gap between proximal and distal nerve stumps sutured into silicon rubber tubes.	Rats that received acupuncture or EA exhibited more mature ultrastructural nerve organization with significantly higher numbers in axon density, blood vessel area, percentage of blood vessel area occupied in total nerve area, than the controls.	Chen et al, Taichung, Taiwan, 2001
60 Wistar rats	Functional rehabilitation of peripheral nerve.	Injury caused by transection of left sciatic nerve; treated with EA.	Compared with control nerve muscle-action potential and motor nerve conduction velocity were better ($p<.01$); single muscle twitch and tetanization of gastronemius muscle were higher ($p<0.05$); conclude EA improves functional rehabilitation of injured peripheral nerve.	Yu et al, Sichuan, 2001

Table 3.1.2. Effect of Acupuncture on Nervous System (cont.)

Subject	Objective	Method	Results	Authors
rats	On nerve growth factor (NGF) and ovarian morphology.	Experimentally induced polycystic ovaries (PCO) by injection of estradiol valerate.	Higer concentrations of NGF were found in ovaries and adrenal glands in rats in the PCO model; after EA concentration of NGF reduced to normal; suggests EA inhibits hyperactivity in sympathetic nervous system.	Stener-Victorin et al, Goteborg, Sweden, 2000
12 healthy volunteers	On sympathetic and parasympathetic activities	Use power spectral analysis: the low and high frequency components of heart rate variability is used to reflect sympathetic and parasympathetic activity.	Ear acupuncture at LU1 increased parasympathetic activity ($p<.05$), no significant change in sympathetic activity, blood pressure, or heart rate;LI4 resulted in significant increase in sympathetic and parasympathetic activity and a significant decrease in heart rate ($p<.05$).E3	Haker et al, Stockholm, Sweden, 2000
humans	On sympathetic nerve activity	2 Hz at LI 4 and LI11; multiunit efferent postganglionic sympathetic nerve activity was recorded with a tungsten microelectrode in muscleof perineal nerve.	EA produced an increase in pain threshold and a transient increase in muscle sympathetic nerve activity.	Knardahl et al, Goteborg, Sweden, 1998

Table 3.1.2. The Effect of Acupuncture on Nervous System (cont.)

Subject	Objective	Method	Results	Authors
rats	On afferent fiber responsible for the suppression of jaw opening reflex (JOR).	Manual acupuncture stimulation of various areas, such as nose, auricle, forepaw, abdomen, hind leg and hind paw.	Results suggested that the capsaicin-sensitive thin afferent fibers (A delta and C afferent fibers) mediated by receptors such as polymodal receptors are activated and participated in the inhibition of JOR by diffuse noxious inhibitory controls.	Okada et al, Kyoto, 1996
54 cases of peripheral nerve injury; 54 control	Treatment by EA.	Changes were observed by electromyography. Sensory and motor improvement were also recorded.	5 cases cured, 26 markedly effective, 19 improved and 4 cases failed, with effective rate of 92.6%. Radial nerve and common pereneal nerve recovered faster than others.	Hao et al, Henan, 1995

Table 3.1.3. Effect of acupuncture on reproductive system

Subject	Objective	Method	Results	Authors
female rats	On aging process of genital system.	EA at BL23	The frequency of neuronal discharges in locus coeruleus (LC) was elevated; the activating rate of LC to neurons in the medial preoptic area of the hypothalamus was increased; EA raised the catecholamine (CA)/5-hydroxytryptamine (5-HT) ratio in the hypothalamus of aged rats so as to delay the aging process.	Zhu et al, Shaanxi, 2000
24 women	On anovulation in women with polycystic ovary syndrome (PCOS).	10-14 EA treatments.	An increase of ovulations in women from 0.15 to 0.66 (p=.004) for 38% of the women.	Stener-victorin et al, Goteborg, Sweden, 2000
16 patients	Treatment of erectile dysfunction.	5 Hz EA on four acupoints for four weeks, twice a week.	15% of ptients experienced an improvement of the quality of erection; 31% reported an increase in their sexual activities.	Kho et al, Nijmegen, Netherlands, 1999

Table 3.1.3. Effect of acupuncture on reproductive system (cont.)

Subject	Objective	Method	Results	Authors
anovulatory patients	Normalization of hypothalamic-pituitary-ovarian axis (HPOA).	EA; clinical observations.	EA might regulate HPOA in several ways: influence some gene expression of the brain; normalizing secretion of some hormones, such as GnRH, LH, and E2.	Chen et al, Shanghai, 1997
rat brain	On the expression of extrogen receptor protein and mRNA.	Radioimmunoassay, RNA dot blot and Northern blot, monoclonal antibody immuno-histochemistry; computer image analysis.	Results show that ovariectomy induced a decrease of blood estradiol (E2) level and increase of the expression of estrogen receptor protein and mRNA. These effects could be reversed by previous treatment of EA of ovariectomized rats; suggested that EA may activate the production of body estrogen.	Chen et al, Shanghai, 1997
44 patients and rats	On hyperplasia of mammary glands and immunological function.	EA treatment of patients and rats with hyperplasia mammary glands.	Decrease of immunological function in hyperplasia of mammary glands of patients and rats. EA enhanced this function. Implies that the inhibitory effect on immunological function exerted by high concentration of E2 was lowered by EA. Hence, immunological function was restored.	Guo et al, Shaanxi, 1996

Table 3.1.3. Effect of acupuncture on reproductive system (cont.)

Subject	Objective	Method	Results	Authors
10 infertile women	On reduction of blood flow impedance in uterine arteries.	Pulsatility index PI<=3.0; EA twice a week for 4 weeks.	The mean PI was significantly reduced after 8th EA ($P<0.001$), and 14 days after treatments ($p<0.001$); skin temperature on the forehead (STFH) incrased significantly during EA, probaly due to a central inhibition of sympathetic activity.	Stener-Victorin et al, Goteborg, Sweden, 1996
18 ovariectomized rats	On nucleolar organizer regions (NOR) of the adrenal cortex.	The number of argyrephil (AgNCR) of 100 cells from zona fasciculata of the adrenal cortex was counted.	The mean of AgNOR of avariectomized rats with EA differed disgnificantly from controls. Suggests that EA promotes the synthesis and secretion of adrenal steroid hormones, the androgen of which will then be transformed into estrogen in other tissues, thus compensating the deficiency of estrogen induced by ovariectomy.	Chen et al, Shanghai, 1994
55 male rats	On testosterone (T), dihydrotestosterone (DTH) in blood plasma	Rats operated on with bilateral testectomy, adrenalectomy; radioimmunoassay was used to determine T and DTH in femoral vein blood.	T from testes and adrenal and DHT rose markedly after EA at ST36. Suggested EA may be used for improving the hypogonadal condition.	Kong et al, Henan, 1991

Table 3.1.3. Effect of acupuncture on reproductive system (cont.)

Subject	Objective	Method	Results	Authors
48 cases of women with malposition of fetus	Correct malposition of fetus.	EA at Zhiyin BL67; another group with moxibustion.	39 cases were corrected with a rate of 81.3%. Sessions of EA were fewer than moxibustion. There is no significant difference of efficacy between the two methods.	Li et al, Changchun 1996
female rabbits	Release of gonadotropin-release hormone (GnRH).	EA, 4Hz, at RN4, RN3, SP6, and EX16.	Ovulation is usually accompanied by pulsatile secretion of gonadotropin, the release of GnRH from mediobasal hypothalamus (MBH); EA increased GnRH significantly from $4.64+/-0.98$ from $2.08+/-0.01$.	Yang et al, Shanghai, 1994
11 cases	EA ovulation induction.	Relationship between radioimmunolo-reactive beta-endorphin (r beta-E) and hand skin temperature (HST).	There is a negative correlation in the decrease of blood r beta-E and increase of HST after EA ($p<0.01$). EA is able to regulate the function of the hypothalamic-pituitary-ovarian axis.	Chen et al, Shanghai, 1991

Table 3.1.3. Effect of acupuncture on reproductive system (cont.)

Subject	Objective	Method	Results	Authors
oligo-menorrheic women	Hormonal status.	Three groups: ovarian hypo-fuction, ovarian hypofunction with hyperandrogenism, ovarian hypofunction with hyperprolactinemia.	Treatment with acupuncture normalized the hormonal status in all 3 groups. The hormonal status of oligomenorrheic patients correlated with conductivity of the skin at biologically active sites.	Parshutin et al, Russia, 1990
60 pregnant women	Induction and inhibition of labor.	EA at special sites of extremities.	48 cases for labor induction; in 32 cases, delivery was achieved, success rate 78%; in 11 of 12 cases of premature labor, all carried pregnancy to term with a success rate of 91.8%.	Tsuei et al, 1977

Table 3.1.4. Effect of Acupuncture on Gastrointestinal System

Subject	Objective	Method	Results	Authors
17 children, constipated	On chronic constipation.	Parents filled in bowel habit questionnaire; basal plasma panopiod level was measured. 10 weekly acupuncture sessions.	The frequency of bowel movement in males increased from 1.4+/-0.6/week to 4.4+/-0.6/week after ten weeks. Females improved from 1.4+/-0.3/week to 5.6+/-1.2/week. Basal panopioid activity was lower before treatment, and returned to normal after treatment.	Broide et al, Zerifin, Israel
25 healthy subjects	Relieving motion sickness and reducing abnormal gastric activity.	Wear commercial Acuband wrist band; use rotating opto-kinetic drum. Recording by electrogastrography.	An Acuband worn on the wrist or forearm decreases the symptoms of motion sickness and the gastric activity that usually accompanies motion sickness.	Stern et al, Penn State U, 2001
26 patients	Postoperative recovery of intestinal function.	Auricular-plaster therapy and acupuncture at ST36.	12 out of 13 cases in the treatment group (92.4%) showed recovery of normal peristalsis within 72 hours of operation; in control group only 46.1% recovered.	Wan, Nanjing, 2000

Table 3.1.4. The Effect of Acupuncture on Gastrointestinal System (cont.)

Subject	Objective	Method	Results	Authors
mice	Reciprocal actions	Orthogonal design was used to observe gastrointestinal peristalsis in normal and atropine-treated mice by EA.	EA has no effect on normal mice; BL20 was orthogonal to ST36; the decreased peristalsis in atropine-treated mice was markedly promoted by EA at ST36; P6 was antagonistic to Bl20 in atropine-treated mice. Reciprocal action among acupoints should be taken into consideration in treatment.	Xu et al, Jiangxi, 1999
7 patients	Irritable bowel syndrome.	Assessment was by a diary card.	Acupuncture seems to be effective in the treatment of irritable bowel syndrome.	Chan et al, 1997
60 patients	Hypotonic effect on gastrointestinal tract.	Comparison of the width of corpus and antrum of the stomach and duodenum by aricular acupuncture.	The effect of auricular acupuncture on the motility and tone of gastrointestinal tract was the same as with the usual drugs.	Borozan et al, Beograd, 1996
mice	On intestinal motility.	The distance of intestinal movement of a carbon solution injected into the stomach.	Intestinal peristalsis was accelerated significantly by acupuncture at the abdomen, but suppressed by moxibustion. The intestinal peristalsis acceleration by vagostigmin was reduced significantly by both acupuncture and moxibustion.	Iwa et al, Kyoto, 1994

Table 3.1.4. Effect of Acupuncture on Gastrointestinal System (cont.)

Subject	Objective	Method	Results	Authors
20 rabbits	On dysrythmia of gastro-colonic electric activity (GEA).	Gastron-colonic disorder by injection of erythromycine (EM); effect of EA on disorder of GEA.	After injection of EM, the frequency and amplitude of GEA were increased. EA could shorten the duration latency of EM effect, decrease the frequency and variation coefficient of GEA. After vagotomy, effect of EA disappeared.	Xu, Hefei, 1994
rats	Reduce the effect of stress on intestinal [tract?]	Three pairs of bipolar electrodes were implanted on antrum and colon; acupuncture at ST36; gastroenteric electric activity (GEA) was measured.	Stress is expressed as a the reduction of frequency and amplitude of slow wave in GEA. Acupuncture at ST36 could reduce the inhibition effect on GEA induced by stress, which was caused by cold immersion.	Xu G, Anhui, 1994
rabbits	On gastric hyperfunction	EA on ST36 and ST44; excitation of lateral hypothalamus area (LHA) by electrostimulation.	EA inhibits hyperactivity of stomach induced by hyper-activity of LHA ($p<.05$); beta-receptor blocking agent propranolol in turn inhibits the effect of EA. EA has an anticholinergic effect through gastric beta receptor, therefore it inhibits the appetite, relieves hunger and reduces body weight in obese people.	Ma et al, Nanjing, 1994

Table 3.1.4. Effect of Acupuncture on Gastrointestinal System (cont.)

Subject	Objective	Method	Results	Authors
rats	On gastric motility in anesthetized rats.	Gastric motility in the pyloric region measured with the balloon method.	The inhibitory gastric response by EA of the abdomen is a reflex response from the spinal cord; its afferent nerve pathway is comprised of abdominal cutaneous and muscle afferent nerves. The afferent nerve pathway is gastric sympathetic nerve. EA on the hind paw is a reflex response from the brain, comprised of cutaneous and muscle afferent nerves, and gastric vagal efferent nerve. Endogenous opioids are not involved.	Sata et al, Tokyo, 1993
57 rats	On gastro-duodenal mucosal lesion and gastro-duodenal electroactivities.	Acupoints at ST36 and BL21; bipolar electrodes implanted on antrum, duodenum and abdomen; stress induced by cold water.	(1) Under stress 63.2% showed area density of lesion to be 15.8-27.7%; 84% showed inhibition of gastro-duodenal electroactivities: amplitude of slow waves was lower, number of fast waves reduced, cycle was prolonged. (2) Under stress and EA, only 16.6% showed lesion area density to be 1.7%; gastro-duodenal electrical changes were hardly seen. Inhibitory states were significantly reduced 54.5%.	Xiang et al, Shanghai, 1993

Table 3.1.4. Effect of Acupuncture on Gastrointestinal System (cont.)

Subject	Objective	Method	Results	Authors
39 patients with abdominal surgery	On intestinal motion and sero-enzyme activity; acupoint ST36, SP6	Time of first excretion; sero-enzyme activity of GPT, GOT, and gamma-GT before and after operation.	Time of first excretion is 57+/-23.94 h as compared with control group's 86.14+/-20.43h (p<.001); after, sero-enzyme activity was raised 2-3 times higher than before. Suggested acupuncture regulated reactivity of organism to trauma and promoted the repair of damaged cell.	Liu et al, Beijing, 1991
20 guinea pigs	On gastroenteric electric activity (GEA) of peripheral vomiting.	EA at ST36; four pairs of electrodes implanted under serosa of antrum, corpus, duodenum and jejunum.	EA could shorten the duration of peripheral vomiting, could reduce the number of fits (p<.001), and could lower the amplitude and frequency of the GEA (p<.05).	Len et al, Anhui, 1991
rats	On beta-endorphin of tracts.	ST36; use radio-immuoassay to measure beta-endorphin(ir) contents in gastrointestine, anterior lobe of the pituitary, and plasma.	ir was significantly elevated in gastroparietal mucous membrane, pyloric mucous membrane, duodenum, jejunum and ileum; but ir in pituitary and plasma was not elevated. Suggested regulation of gastrointestinal tract through the endogenous opiate peptides in gastrointestine.	Yang et al, 1989

Table 3.1.4. Effect of acupuncture on Gastrointestinal System (cont.)

Subject	Objective	Method	Results	Authors
220 patients with purulent peritonitis	To restore motor-evacuatory function of stomach and intestine.	Acupuncture	Positive effect was noted in all patients with local peritonitis after single procedure. In 100 patients with diffuse peritonitis, no effect in 20 patients.	Kabanov et al, 1989

Table 3.1.5. Effect of Acupuncture on Urinary System

Subject	Objective	Method	Results	Authors
anesthetized rats	On sympathetic nervous and hyperactive bladder.	EA at L14; response of rhythmic micturation, excretion, blood pressure, renal sympathetic and pelvic nerve activity.	Different frequencies (2Hz and 20Hz) selectively activate distinctive sympathetic, but not parasympathetic nervous system. EA at L14 may ameliorate hyperactive bladder.	Lin et al, Taipei, 1998
rats; spontaneously active	On blood pressure and renal function.	Moxibustion at BL15 and BL27.	Urine volume increased for BL15, but decreased for BL 27. Systolic blood pressure decreased for BL15, but stayed the same for BL27. Aldosterone and renin increased, but atarial natriuretic peptide decreased for BL15. Both aldosterone and atrial natriuretic peptide increased for BL27. So they both regulate renal function and secretion of hormone with body fluid metabolism.	Lee et al, Iri, Korea, 1997
11 patients with overactive bladder	On overactive bladder.	Bilateral BL33 for average 7 times.	Urge incontinence was controlled completely in 5 and partially in 2 of 9 patients. Uninhibited contraction disappeared in 6 patients. Uninhibited contraction disappeared in 6 patients after treatment. Acupuncture induced an increase of maximum bladder capacity ($p<.01$).	Kitakoji et al, Mejii College, 1995

Table 3.1.5. Effect of Acupuncture on Urinary System (cont.)

Subject	Objective	Method	Results	Authors
rat	On urinary bladder	Electric discharges of the post-pelvic ganglionic vesical nerve fibers were recorded.	The changes of the pressure-volume of urinary bladder are controlled not only by the wall of the urinary bladder but also by the regulation of parasympathetic nerve. EA regulates urinary bladder through nerves.	Ben et al, Beijing, 1995
rabbits	On renal blood flow (RBF)	EA at BL23; glycerol-induced renal ischemia and renal neurotomy. Measured RBL by hydrogen gas clearance method.	(1) RBL was decreased by EA in normal and induced renal ischemia rabbits. (2) After renal neurotomy, RBT is increased by EA. This suggested that EA affects both renal nerve and body fluid.	Xu et al, Anhui, 1995
25 children with enuresis	On neurogenic bladder dysfunction.	Urodynamics of the lower urinary tract was evaluated.	In 17 cases acupuncture was beneficial. Follow-up showed detrusor-stabilizing effect in patients with neurogenic bladder dysfunctions. Success depended on patients' mental and emotional status, concurrent abnormalities and accuracy in observing practitioner's recommendations.	Kachan et al, Russia, 1993

Table 3.1.5. Effect of Acupuncture on Urinary System (cont.)

Subject	Objective	Method	Results	Authors
male rabbits	Effect on partially denervated bladder.	Postganglionic fibers from bilateral pelvic ganglia to the bladder were excised partially and resulted in reversible retention of urine; auriculopoint "sub-cortes" and BL23.	After operation, pressure-volume curve ascended a step after every injection of water instead of a peak, as before operation; the bladder volumen and the residual urine increased after operation. In EA group the residual urine decreased more evidently.	Ben et al, Beijing, 1993.
rabbits	Effect of EA at K13 on renal blood flow RBL.	The RBF was measured by hydrogen clearance method.	(1) After EA, RBL increased. (2) Under the condition of renal ischemia, the thromboxane A2 (TXA2) increased, and prostacyclin (PG12) decreased. (3) Then EA decreased TXQ2 and increased PG12.	Xu et al, Hefei, 1993
anesthetized rats	On rhythmic micturation contraction (RMC) of urinary bladder.	Infused saline until RMC resulted from rhythmic burst discharges of the vesical pelvic efferent nerves.	Inhibition of the RMC by acupuncture-like stimulation of the perineal area is a reflex response characterized by segmental organization. The afferent arcs of the reflex are both pelvic and pudendal nerve brances innervating the perineal skin and underlying muscles, while the efferent arcs are pelvic nerve branches innervating the urinary bladder.	Sato, et al, Tokyo, 1992

Table 3.1.6. Effect of Acupuncture on the Liver

Subject	Objective	Method	Results	Authors
36 rats	On treatment of acute liver damage.	Carbon tetrachloride induced liver injury; acupoints ST35 and L13.	Rats treated with CCl4 had higher levels of serum glutamate-oxalate-transaminase and serum glutamate-pyruvate-transaminase. After acupuncture biochemical and pathological parameters of liver injury were significantly reduced.	Liu et al, Taoyuan, Taiwan, 2001
33 rats	Protection against hyperbrillirubinemia and cholangitis.	Oral adminsitration of alphanapthylisothiocyanate (ANIT)	Rats treated with ANIT exhibited elevations in brillirubin, SGOT and SGPT as well as cholangitis. Rats treated with ANIT and acupuncture had significantly reduced biochemical and morphological parameters of liver injury.	Lin et al, Taichung, Taiwan, 1995
rats	Effect of EA on hepatic liver.	Acute (7Hz, 0.75V); chronic (4Hz, 0.75V) EA for 21 days; cellular and subcellular levels. Acupoints: BL23, BL25, ST36.	After chronic treatment, (a) protein, RNA, phospholipid and cholesterol of liver increased significantly. (b) Liver microsomal G-6-pase activity increased significantly. (c) Microsomal lipid peroxidation value decreased. (d) Lipase activity increased. After acute treatment, (e) phospholipid and cholesterol of liver increased significantly, (g) liver microsomallipid peroxidation value decreased, (h) GPT and lipase activity increased.	Chakrabarti et al, 1983

Table 3.1.8. Effect of Acupuncture on Stomach

Subject	Objective	Method	Results	Authors
rats	On injured gastric mucosa.	Observe changes in endothelium-derived factors – nitric oxide and endothelin (ET); injured gastric mucosa by ethanol. EA at ST36.	The protective effect of EA on injury of gastric mucosa was due mainly to nitric oxide.	Sun et al, Hefei, 1999
rats	On protecting against stress peptic ulcer.	Stress-induced gastric ulcer models were established by immersion of restrained rats in water.	EA was able to protect stressed rats from stress-induced peptic ulcers, probably by enhancing gastric mucosal barrier, stabilizing gastric mast cells and inhibiting the gastrin levels in gastric mucos.	Shen et al, Hefei, 1995
108 rats	On gastric ulcer.	Experimental gastric ulcer; body weight, ulcer area, plasma SOD and LPO of each rat were observed.	10 minutes EA treatment using thinner needles every other day was the best method.	Fu et al, Guangzhou, 1995
rats	On protecting stress peptic ulcer.	Electroacupuncture group and control group.	Synthesis and decomposition of central and peripheral 5-HT in EA rats was inhibited; more NE levels in three brain regions and blood were seen; higher levels of DA in gastric tissue and blood were seen. Suggested EA was connected with monoamine neurotransmitters of center and periphery.	Shen et al, Hefei, 1994

Table 3.1.8. Effect of Acupuncture on Stomach (cont.)

Subject	Objective	Method	Results	Authors
rats, 30 pairs	Protective effect on gastric mucosa by EA.	Changes of neurotransmitters were studied by histochemical methods of cholinesterase and catecholamine.	There was a protective effect on gastric mucosa. Both cholinergic and adrenergic nerves underwent significant inhibition.	Pan et al, Hefei, 1990
male white rats	On signs of emotional stress caused by pain.	EA on the point analog Da-Djuy ; compared with the effect of diazepam.	EA abolished consequences of pain-induced acute emotional stress. Resulted in a significant reduction of gastric erosions and in emotional reactivity, increased aggressiveness when exposed to stress.	Andreev et al, Russia, 1981
9 healthy Chinese	On gastric myoelec-trical activity.	Gastric myoelectrical activity was recorded using surface electrogastrography (EGG).	EA significantly increased the percentage of regular slow waves. May enhance the regularity of gastric myoelectrical activity and may be an option for teatment of gastric dysrhythmia.	Lin, U of Oklahoma 1997

Table 3.1.9. Effect of Acupuncture on Lung

Subject	Objective	Method	Results	Authors
111 patients with obstructive chronic lung diseases	On pulmonary vascular resistance.	Laser puncture; 10 sessions, 890 nm wave length, 1500 Hz and 2mW mean radiation rate.	Improved bronchial potency, enhanced bronchial sensitivity to sympathominetics, reduced systolic pressure in the pulmonary artery.	Zamotaev et al, Russia, 1991
30 patients	On obstructive forms of chronic nonspecific lung diseases.	Pulmonary volumes pattern, changes in bronchial permeability and partial oxygen pressure in capillary blood.	Laser puncture was found to promote the benefit of standard acupuncture treatment, to improve bronchial permeability, enhance oxygenation of arterial blood, reduce dyspnea and cough, to stimulate expectoration, normalize sleep and appetite.	Zamotaev et al, Russia, 1990
10 patients with Chronic obstructive pulmonary diseases (COPD)	On quality of life, mouth occlusion pressures, and lung function.	Pulmonary functiont tests and interview with questionnaire. Two weeks of treatment of 7 verum acupuncture.	Improvement of large magnitude in quality of life, and lower demand on the respiratory pump.	Neumeister et al, Ruhr Universitat, 1999
12 matched pairs of patients with COPD	For disabling breathlessness.	Three weeks' treatment of acupuncture group and placebo group.	Showed significantly greater benefit in terms of subjective scores of breathlessness and six-minute walking distance.	Jobst et al, 1986

Table 3.2.1. Effect of Acupuncture on Cardiac Disorders

Subject	Objective	Method	Results	Authors
A. Animals				
open chest dogs	Benefit of EA at PC6 at 40Hz.	Open chest dogs under pentobarbital and fentanyl anesthesia. Measure mean arterial pressure, end-diastolic volume, heart rate, stroke volume, cardiac output, end-systolic pressure.	All these parameters decreased by 5 to 10% over 1.5 hours, without EA. EA at PC6 increased these cardiovascular variables by 10-15%, especially end-systolic elastance (EES) by 40%	Syuu et al, Japan, 2001
rabbits with acute myocardial ischmeia (AMI)	On myocardial microcirculation.	EA at PC6; observe myocardial microcirculation and electrical activities by vascular casting method and taking monophasic action potential.	EA can relieve arteriolospasm, inhibit extreme dilatation of blood capillaries, modulate imbalance of microvasomotion of the corony artery, improve myocardial blood supply, and promote normalization of electrical activities of ischemia myocardium.	Cao et al, Beijing, 1998
rabbit	On H^+ concentration.	Arrhythmia induced by aconitine.	There was a specific change of H^+ concentration in PC6 as well as along the heart meridian and the pericaridium meridian.	Wang et al, Tianjin, 1996

Table 3.2.1. Effect of Acupuncture on Cardiac Disorders (cont.)

Subject	Objective	Method	Results	Authors
44 rabbits	Influence of neuron excitation and inhibition of rVLM on EA at PC6.	Rabbits were anesthetized with a mixture of urethane and chloralose; microinjection of glutamate and glycine.	(1) EA could signifcantly promote the recovery of ECG ST-segments, mean blood pressure, left ventricular pressuer (LVP), and the maximum rising rate of LVP. (2) Glutamate could enhance the effects of EA. (3) Glycine weakened the effects of EA. Conclusion: rostal ventrolaterenal medulla (rVLM) participated in the regulatory effect of EA on heart.	Liu et al, Beijing, 1996
40 rabbits	Influence of noradrenaline (NA) on the effect of EA.	Thoracic spinal subarachnoid microinjection of NA; EA at PC6.	EA at PC6 could significantly accelerate the recovery of ST-segment and T-wave of electrocardiogram induced by AMI. Injection of NA strengthened the effect of EA. Suggested alpha-receptors if ontra-thoracic spinal cord participate in the action of EA, ameliorating AMI.	Liu et al, Beijing, 1995
rats	On threshold of ventricular fibrillation threshold.	EA at PC6 Neiguan and Lingdao on anesthetized Wistar rats.	The value of ventricular fibrillation threshold (VFT) was increased significantly in 30 min. of electroacupuncture (p<0.01).	Zhang et al, Hubei, 1999

Table 3.2.1. Effect of Acupuncture on Cardiac Disorders (cont.)

Subject	Objective	Method	Results	Authors
30 mongrel dogs	On myocardial oxygen metabolism and pH of blood.	Acute myocardial ischemia was produced by reducing blood supply. EA at 1-20 Hz and 5 volts.	Results indicated that EA could reduce oxygen consumption of ischemic myocardium, prevent the decrease of pH of coronary sinus blood. Thus, myocardial cell acidosis was prevented, myocardial contractile force was strengthened.	Gao et al, Beijing, 1992
rabbits	Role of amygdaloid nucleus.	EA at Neiguan PC6; and acute myocardial ischemia (AMI) in rabbits.	EA could regulate the changes in the discharge of single units in AMYG caused by AMI, that is, EA inhibited the increased frequency and reversed the decreased frequency. Suggested AMYG is an important link in the heart and the acupoint PC6.	Lai et al, Beijing, China, 1991
rabbits with AMI	On the effective refractory period.	Induce incomplete ischemia by pulling the ventricular branch of the left coronary artery of rabbits.	EA at PC6 could inhibit reduction of effective refractoriness of ischemic myocardium and improve the dispersion of refractoriness markedly, which would be beneficial to synchronization of the myocardial excitatory state.	Cao et al, Beijing, 1993
15 rats	On PC6 – heart short reflex.	AMI induced by pituitrin; dorsal roots were removed from spinal cord segments, cervical 6, C6-thoracic (T1).	Acupuncture at PC6 can influence the superficial electrical resistance of acupoint PC6, improve the ECG, and increase the heart rate. PC6 at ST36 has no such effects. It suggested that PC6 and hart may be connected through a short reflex.	Wang et al, wuhu, China, 1991

Table 3.2.1. Effect of Acupuncture on Cardiac Disorders (cont.)

Subject	Objective	Method	Results	Authors
rabbits	On PO-AH and PHA.	The activity of PHA neurons was recorded by using glass microelectrodes.	The AMI-induced PHA neurons activity could be changed by EA at PC6; the effect of EA could be enhanced in some PHA neurons by PO-AH stimulation under EA. Suggested that PO-AH and PHA possess some coordination effect in the regulation of EA at PC6 on the functional activity of the heart.	Liu et al, Beijing, China, 1990
B. Patients				
48 healthy persons	On cardiopulmonary function.	Effects are measured with gas analysis; acupoints are PC6 and ST36.	EA can decrease resting heart rate and carbon dioxide production ($p<0.05$), plus lower the metabolic rate.	Lin et al, Taipei, Taiwan, 1996
20 coronary heart disease patients.	On heart rate variability (HRV).	EA at PC 6; frequency domain analysis of HRV.	In EA group the low frequency/high frequency change was marked ($p<0.05$). Suggested that acupuncture could regulate and improve HRV in coronary heart disease patients.	Shi et al, Harbin, China, 1995

Table 3.2.1. Effect of Acupuncture on Cardiac Disorders (cont.)

Subject	Objective	Method	Results	Authors
patients with coronary heart disease	On different times of day.	Measure left ventricular function.	EA which was performed early, 7-9am, improved the left ventricular function by shortening of PEPI and decrease of PEPI/VETI ratio. EA which was performed late, 7-9 pm, prolonged PEPI and raised PEPI/LVETI ratio, suggesting impairment of left ventricular function.	Li et al, Shanghai, 1994
100 cases, coronary heart disease	On relation between PC meridian and cardiac function.	Electrocardiogram, cardiophonogram and rheocardiogram were recorded; 4 acupoints: PC6, PC4, PC3, PC2 and two nonacupoints on PC.	8 indices: PEP, LVET, P/L, HI, SV, CO, ST segment and T wave of ECG. These indices all changed conspicuously after acupuncture. 737 observations were made.	You et al, Fuzhou, China, 1993
300 coronary heart disease patients	On propagated sensation along meridian.	Acupuncturing at PC6 with same manipulation by identical doctor.	Higher rate of PSM appearance and better acupuncture effects were observed in patients with compound sensations, such as sourness-distension and distention-numbness.	You, China, 1992

Table 3.2.1. Effect of Acupuncture on Cardiac Disorders (cont.)

Subject	Objectives	Method	Results	Authors
21 patients	On angina pectoris.	Randomized crossover study, three times a week, at PC6, HT5, B15, B20, ST36, for four weeks. Previous anitaginal treatment remained unchanged.	During acupuncture the number of angina attacks per week was reduced from 10.6 to 6.1, compared with placebo (p<0.01). Intensity of pain at maximal work-load decreased from 1.4 to 0.8 (p<0.01). ST-segment depressions decreased from 1.03 to 0.71 mm (p<0.01). Questionnaire confirmed improved feeling of well-being.	Richter et al, Sweden, 1991

Table 3.2.2. Effect of Acupuncture on Arthritis

Subject	Objective	Method	Results	Authors
A. Rats				
adjuvant arthritis rats	To produce antinociceptive and anti-inflammatory effect.	Inject bee venom into Zusanli ST36 of Freund's adjuvant-induced rheumatoid arthritis.	The injection of water soluble fraction dramatically inhibits paw edema and radiological change, i.e., new bone proliferation and soft tissue swelling.	Kwon et al, Seoul, Korea, 2002
adjuvant arthritis rats	Effect of intrathecal injection of somato-statin and EA.	C-fos protein immunoreactivity (FLI) was detected by immunohistochemistry.	Pathological pain following arthritis activated pain-sensitive neurons (PSN) and evoked c-fos expression in spinal cord. Somatostatin and EA suppressed activities of PSM, producing the effect of analgesia.	Ruan et al, Chongqing, 1997
adjuvant arthritis rats	Role of opiate-like substance in acupuncture analgesia (AA).	Needle at Huantiao, naloxone block the effect of AA.	Acupuncture may enhance the release of peripheral opiate-like substance which can act at the sensitized opiate receptors, leading to more potent analgesic effect in the inflamed area.	Zhu et al, Beijing, 1993

Table 3.2.2. Effect of Acupuncture on Arthritis (cont.)

Subject	Objective	Method	Results	Authors
adjuvant arthritis rats	On cotical and hippocampal (HPC) EEG.	The HPC EEG were recorded; behavior and local inflammation observed.	Desynchronization of ECoG and HPC EEG in the AA rats. The delta waves were decreased and beta waves increased significantly, suggesting cortex and hippocampus participate in the modulating action of AA.	Lou et al, Guangzhou, 1992
B. Patients				
32 patients; osteoarthritis of the hip	Comparison of acupuncture with advice and exercise.	Randomized controlled trial; patients were assessed for pain by WOMAC.	There is significantly more improvement in treated group than in control group immediately post-treatment (p=0.002). This was maintained at the eight-week follow-up (p=0.03).	Haslam, Swindon U
44 patients with advanced osteoarthritis of the knee.	Unilateral verus bilateral acupuncture.	Acupuoints are SP9, SP10; ST34, ST36. Blinded observer assessed knees before and at the end of two and six months.	Significant reduction in symptoms in both kinds of treatment. The unilateral acupunctue is as effective as bilateral acupunctue in increasing function and reducing the pain associated with OA of the knee.	Tillu et al, Bedford, UK, 2001

Table 3.2.2. Effect of Acupuncture on Arthritis (cont.)

Subject	Objective	Method	Results	Authors
73 persons; osteoarthritis of the knee	Adjunctive therapy	Twice a week for 8 weeks; patients self-scored on WOMAC and Lequesne Algofunctional index.	Acupuncture may be best used early in the treatment plan, with a methodical decrease in frequency in treatment once the acute teatment period is completed, to avoid a rebound effect. Demographic and medical history data were not mediating variables.	Singh et al; SCU, CA, USA, 2001
67 patients; osteoarthritis of the hip	Non-specific effects of acupunctue.	Pain(VAX), functional impairment (hip score), activity in daily life (ADL) and overall satisfaction.	For all parameters there was a significant improvement versus baseline 2 weeks and 2 months following teatment. There were no differences between needling at acupoints and at points near them at LA to L5 dermatomes.	Fink et al, Hannover, Germany, 2001
40 patients; osteoarthritis of the hip and knee	Combined treatment of acupuncture and drugs.	Subjects received non-steroidal anti-inflammatory drugs with intra-articular administration of glucocorticosteroids and inhibitors of proteases.	Indicators of articular pains and tenderness of joints in palpation got significantly lower by the end of the course of teatment in those patients (20) receiving additional acupuncture treatment.	Zherebkin, Russia, 1998

Table 3.2.2. Effect of Acupuncture on Arthritis (cont.)

Subject	Objective	Method	Results	Authors
116 cases with gonococcal arthritis	Acupuncture.	Du14, LI11, ST 36 acupoints and points along meridian indicated by the affected joint.	Of the 116 cases, 74 were cured, 13 were markedly helped, 11 improved and 11 failed. The effective rate was 84.5%	Wang, Gansu, China, 1996
31 patients; rat rheumatoid arthritis (RA)	Relationship between arthritis and blood stasis.	Most patients had the signs of blood stasis and abnormal hemorheology; similar in rats.	(1) There is a close relationship between RA and blood stasis. (2) Acupuncture promotes blood circulation to remove blood stasis. (3) Examination of hemorheology may set standard for RA and curative effect.	Sun, Shanghai, 1995
54 cases of rheumatoid arthritis	Effect of acupuncture and point injection with zhuifengsu.	Changes in cellular and humoral immunity and other parameters in pheripheral blood.	Good clinical effects were observed with 100% rate. The NK activity and IL-2 value in RA patients were found to be lower than those of normal. Both increased after treatment (p=0.01). Suggested treatment exerts a regulatory effect on cellular immunological function.	Liu et al, Beijing, 1994

Table 3.2.2. Effect of Acupuncture on Arthritis (cont.)

Subject	Objective	Method	Results	Author
29 patients with 42 osteoarthritic knees	Long-term study.	Randomized; 49 weeks.	Results were significantly better in those who had not been ill for a long time. It was possible to maintain the improvements. Acupuncture can ease discomfort, and seven patients have responded so well that they do not want an operation.	Christensen et al, Denmark, 1992
35 patients; gonarthritic pain	Acupuncture	Subjective effectiveness of the treatment, using a standard method on the knee.	Patients reported an explicit improvement of their ailments. We unreservedly recommend this program	Zwolfer et al, Vienna, Austria, 1992
41 patients; 19 healthy persons	Effect of acupuncture and moxibustion.	41 patients were divided into warming needle and point injection group at random.	The results showed that IL-2 levels in two RA groups before were lower than healthy persons (p=0.05). After treatment, levels inceased considerably in two RA groups (p=0.01); acupuncture and moxibustion influences neuroendocrine system to improve IL-2 production.	Xiao et al, Beijing, 1992

Table 3.2.3. Effect of Acupuncture on Allergies (summary of 16 studies)

Subject	Objective	Method	Results	Authors
209 cases allergic asthma	Therapeutic effect of drug acupoint.	Comparison of crude herb moxibustion and drug acupoint application.	Short term total effective rate in the group of drug acupoint application was higher than that in crude herb moxibustion.	Lai et al Guanzhou, China, 2001
192 patients; bronchial asthma	Correction of autonomic system disorders.	Acupuncture was added to machines in treatment; assessed by changes in rythmogram, respiratory function and psychological status.	It was found that corporal acupuncture not only improves bronchial permeability but also reduces psychovegetative disorders. The effect of acupuncture was not related to that of placebo	Rakhov et al, Russia, 2001
20 patients; 18 control	Immunocodulatory effects on asthma.	randomized control study; peripheral blood parameters: eosoniphils, lymphocyte, cytokines.	More patients indicated improvement in well-being ($p=0/049$) after acupuncture treatment; CD4+cells ($p=0.014$) increased significantly. IL06 (0.026) and IL-10 ($p=0.001$) decreased and IL-8 ($p=0.05$) rose significantly; lymphocyte proliferation rate rose significantly ($p=0.035$); no. of eosinophils decreased after acupuncture.	Joos et al, Heidelberg, Germany
25 cases of bronchial asthma	Hormone dependent.	The dose of cortisone was decreased by 2 mg every 10 days.	The symptoms in most patients were markedly improved after 15 treatments. Treatment should last 30 sessions for 3 months. The markedly effective rate is 56%, and total effective rate is 96%.	Hu, Beijing, 1998

Table 3.2.3. Effect of Acupuncture on Allergies (cont.)

Subject	Objective	Method	Results	Authors
217 cases	On chronic bronchitis and asthma.	Combining electric stimulation with topical application of drug on acupoints.	The combined therapy was superior to unitary therapy (p<0.05). It also had a good curative effect in both short and long terms.	Chen et al, Beijing, 2000
419 cases	Allergic rhinitis with asthma.	A combined desensitizing therapy using acupoints of head and the upper back extract of allergens.	Differences between lymphocyte transformation, acdophil cell count, IgA, IgG, E-rosetter between treatment group and control group are significant. After 3 years, 68.73% of patients had a curative effect, and 29.12% became better.	Zhou et al, Suzhou, China, 1997
66 randomized patients	On mild bronchial asthma.	The changes in peak flow variability served as criteria. Spirometric analyses and markers in blood and sputum were used.	There was a consistent, positive yet unspecified effect of needling in both treated groups, which can be interpreted as a trend in affecting the asthmatic condition.	Medici, Germany, 1999
69 patients and guinea pigs	On anaphylactic asthma.	Zusanli (ST36) point immuno-therapy (ZPIT).	The patients' total IgE level was reduced, anti-acarid IgE was lowered, SIgA level was raised, the absolute eosinophilic granulocyte level dropped and function recovered. The curative effect was much better than for conventional desensitization therapy.	Chen et al, Hangzhou, China, 1996

Table 3.2.3. Effect of Acupuncture on Allergies (cont.)

Subject	Objective	Method	Results	Authors
guinea pigs	Ear acupuncture on beta-adrenoreceptor in lung tissues.	The maximal binding volume (Bmax) was determined by radiological ligand binding analysis.	(1) The Bmax of beta adrenoreceptor was significantly lower in asthma group (52.4+/-20.1 fmol/mg protein). (2) The Bmax of the group receiving ear acupuncture was similar to healthy group (84.5+/-35.1 fmol/mg) with $p<0.02$.	Chen et al, Beijing, 1996
20 patients with asthma	Effect on immuno-globulins in patients.	Immunoglobulins in patients are measured.	IgG increased ($p<0.01$), IgM and IgE decreased ($p<0.01$); IgA did not change ($p>0.05$). Results indicate that acupuncture exerts modulation action on immunoglobulins of the human body.	Guan et al, Yunan, China, 1995
94 patients	On bronchial asthma.	Neurogenic, humoral and bioenergetic responses; 241 parameters.	These responses proceeded according to adaptation laws and resulted in bronchial hyper-reactivity. Eosinophilic inflammation in the bronchi diminished acupuncture efficacy.	Aleksandrova et al, Russia, 1995
asthmatic patients	On mucosal secretory IgA.	Ig A and IgE are measured.	SIgA and total IgA in saliva, in nasal secretions and IgE in sera were significantly decreased ($p<0.02$, 0.02, 0.001 respectively) after treatment with acupuncture. It suggested attacks of allergic asthma could be effectively inhibited by acupuncture.	Yang et al, Shanghai, 1995

Table 3.2.3. Effect of Acupuncture on Allergies (cont.)

Subject	Objective	Method	Results	Authors
141 type I allergic diseases	Curative effect.	Comparative study of acupuncture and desensitization therapies.	The curative effect was higher in the acupuncture group than in the desensitization group in allergic asthma allergic rhinitis and chronic urticaria.	Lai, Guangzhou, 1993
17 patients	Asthma bronchiale.	Standard questionnaire half a year after acupuncture treatment.	Over 70% of patients reported a significant improvement of their ailments after ten weeks of treatment and half a year later.	Zwolfer et al, U of Vienna, 1993
111 patients	Laser acupuncture on pulmonary vascular resistance.	10 sessions; laser with 890 nm wavelength, 1500 Hz, 2mW.	Positive response evident from improved bronchial potency, enhanced bronchial sensitivity to sympthomimetrics, reduced systolic pressure in the pulmonary artery.	Zanitaev et al, Russia, 1991
100 patients	Bacterial asthma.	Standard acupuncture, micro-acupuncture, and by combined use of modalities.	Highest efficiency for the combined treatment, the least for microacupuncture.	Oivin et al, Russia, 1989

Table 3.2.4. Effect of Acupuncture on Cancer Patients

Subject	Objective	Method	Results	Authors
A. Animals				
sarcoma S180, ascitic mice	On erythrocytic immunity.	Moxa-cone moxibustion at Guanyan RN4.	Moxibustion could significantly increase the decreased erythrocytic C3b receptor rosette forming rate, lower the decreased erythrocytic immuno-complex rosette forming rate ($p<0.05$, and 0.01), increase the decreased activity of erythrocytic immuno-accelerative factor, and reduce the increased activity of erythrocytic immuno-suppressive factor ($p<0.05$). In summary, moxibustion strengthens erythrocytic immunity and promotes its regulative function.	Wu et al, Chengdu, China, 2001
rats	On hyperplasia of mammary glands.	Electro-acupuncture therapy on immunological function and E2 in hyperplasia of mammary glands.	There was decrease of immunological function in hyperplasia of mammary glands, which could be remarkably enhanced by EA in rats and in patients. The results indicate EA may reduce the incidence of breast carcinoma.	Guo et al, Shaanxi, China, 1996

Table 3.2.4. Effect of Acupuncture on Cancer Patients (cont.)

Subject	Objective	Method	Results	Authors
rats	On transplanted mammary cancer.	NK cell activity and T-lymphocyte positive rate of ANAE, and lymphocyte transformation were increased ($p<0.01$).	There were marked differences in pathological section ($p<0.02$), adnoid structure, lymphocytic infiltration, and tumor volume ($p<0.01$) between treatment and control group. Acupuncture probably inhibits growth of mammary cancer and reduces its malignancy.	Liu et al, Shijiazhuang, China, 1995
tumor-bearing mice	Regulation on beta-END.	Moxibustion on Guanyuan RN4.	It stimulated the secretion of beta-END from the pituitary and the adrenal gland, increased the level of serum beta-END significantly and kept the high level for quite a long time. It probably caused a constant release of beta-END.	Zhai et al, Shanghai, 1994
B. Patients				
cancer patients at old age	To increase quality of life.	Laser auriculo-acupuncture with psychotherapy.	Results of treatment: 74% with good effect, 15% with satisfactory effect, 11% without effect. Patients with advanced stages of cancer had a satisfactory effect in only 40% of cases.	Blinov et al, St. Petersburg, Russia, 2002

Table 3.2.4. Effect of Acupuncture on Cancer Patients (cont.)

Subject	Objective	Method	Results	Authors
123 patients, 823 visits	Intergration of acupuncture into oncology clinic.	Questionnaire was administered by phone to 89% of patients, mean of five acupuncture visits.	Most patients (60%) showed at least 30% improvement in their symptoms. One third of patients had no change in severity of symptoms; 86% of patients considered it important to continue acupuncture. Conclusion: acupuncture may control symptoms for cancer patients.	Johnstone et al, San Diego, USA, 2002
50 patients, 318 treatments	Xerostomia after radiation therapy.	Eight needles: three in each finger and one in index finger. Response measured by xerostomia inventory (XI).	Response (defined as improvement of 10% or better over baseline XI values) occurred in 35 patients (70%). Conclusion: acupuncture palliates xerostomiaform for many patients. Recommended thee to four weekly sessions followed by monthly sessions.	Johnstone et al, San Diego, USA, 2002
104 patients	Control of chemo-therapy-induced emesis.	Randomized controlled trial; on classic antiemetic acupoints once daily for 5 days.	The number of emesis episodes occuring during the 5 days was lower for patients with EA ($p<0.00$) than pharmacotherapy group. The difference became insignificant during the 9 day follow-up period.	Shen et al, NIH, Bethesda, USA, 2001

Table 3.2.4. Effect of Acupuncture on Cancer Patients (cont.)

Subject	Objective	Method	Results	Authors
20 patients	Suffering from xerostomia.	Treated for 5 weeks with 10 acupuncture treatments.	Acupuncture had a dramatic effect on xerostomia, and dysphagia and articulation after 5 treatments. Release of neuropeptides that stimulate the salivary glands and increased blood flow are possible explanations.	Rydholm et al, Sweden, 1999
48 patients; mammary cancer	Pain relief and movement improvement.	Patients with mammary cancer after ablation on auxillary lymphadenectomy were treated with acupuncture.	By measuring abduction angle with and without pain. Authors conclude that acupuncture seems to be effective in relieving pain and improving arm movement. The choice of acupoints was important in achieving this improvement.	He et al, U of Saarland, Germany, 1999
48 cases of stomach carcinoma	Analgesia effect on pain by acupuncture.	Including filiform needle group and point injection group.	Both groups had markedly effective rates, which were superior to group taking Western medicine. The toxic and side effects from chemotherapy were prevented; indices of blood rheology were improved, and lowered CU-Zn-SOD activity in erythrocytes was increased; quality of life improved.	Dang et al, Chengdu, China, 1998

Table 3.2.4. Effect of Acupuncture on Cancer Patients (cont.)

Subject	Objective	Method	Results	Authors
7 men with prostatic carcinoma	Treatment of vasomotor symptoms.	Acupuncture twice weekly for two weeks and once a week for ten weeks, after castration therapy.	6 out of 7 completed treatment, and all had substantial decrease in the number of hot flashes. Three months afterwards the number of flashes was 50% lower than before therapy.	Hammer et al, Sweden, 1999
62 cases of stomach cancer & colorectal cancer	Q-wave mm micro-wave irradiation on peripheral white blood cells.	Ten days after operation, irradiation started, either before or after chemotherapy.	The total effective rate was 77.4%; the effective rate for the group receiving microwave irradiation before chemotherapy was significantly higher. Irradiation could promote the hematopoietic function of bone marrow.	Wu et al, Fujian, China, 1997

Table 3.2.4. Effect of Acupuncture on Cancer Patients (cont.)

Subject	Objective	Method	Results	Authors
20 patients	Relief of cancer-related breathlessness.	Sternal and LI4 acupoints were used; measured pulse, respiratory rate, oxygen saturation, patients' scale of breathlessness, anxiety and relaxation.	Seventy per cent (14/20) reported marked symptomatic benefit from treatment; significant changes in patients' scores on breathlessness, relaxation and anxiety. There was significant reduction in respiratory rate ($p<0.02$).	Filshie et al, Surrey, UK, 1996
40 patients with malignant tumors	On immunomodulation.	Changes of T lymphocyte subsets (CD3+, CD4+, CD8+) soluble interleukin-2 (SIL-2R).	Results showed acupuncture has the effect of enhancing the cellular immunity of patients with malignant tumors. Percentage of CD3+, CD4+, and ratios of CD4+/CD8+ and the level of beta-EP was increased, and level of SI-2R was decreased from treatment ($p<0.01$).	Wu et al, Chengdu, China, 1996.
286 cases of bone metastasis of cancer	Treatment of pain.	With analgesic decoction of herbal drugs and low frequency EA.	The total effective rate of reducing pain was 74.2% for 212 cases; the average analgesic effect lasted on the average of 3.6 hours. The rate of lymphocyte transformation rose from 45-76% to 57%-96% after treatment ($p<0.001$).	Guo et al, Shaanxi, China

Table 3.2.4. Effect of Acupuncture on Cancer Patients (cont.)

Subject	Objective	Method	Results	Authors
45 patients with malignant tumors	On interleukin-2 and natural killer (IL-2 and NK).	Acupoints ST36, LI11 and RN6; randomized double blind; daily treatment for 10 days.	The IL-2 and NK cells were lower than in normal patients with malignant tumors. There was a significant increase after treatment (P<0.01). Acupuncture may enhance immune function of patients.	Wu et al, Chengdu, China, 1994
16 males with primary liver cancer.	Relieving effect of acupuncture, herbs and morphine.	Combined treatment of herbs (A), ear-acupuncture (B), and epidural morphine (C) to relieve postoperative pain and abdominal distension.	The A, B, and C reduced narcotics 650, 450, and 550 mg respectively when compared with placebo. A and B minimized abdominal distension and urinary retention, while C prolonged them. A and B accelerated restoration of bowel peristalsis (p<0.05).	Li et al, Shanghai, 1994

Table 3.2.4. Effect of Acupuncture on Cancer Patients (cont.)

Subject	Objective	Method	Results	Authors
51 patients, 48 healthy adults	On T-lymphocytes of patients with malignant neoplasm.	Divided 51 into two groups: one with treatment and the other without treatment. Healthy adults served as control.	The percentage of OKT3+, OKT4+, and OKT8+ cells in 51 patients were lower than those of normal adults. After treatment these three percentages were obviously higher than those before treatment. More effect on OKT4+ than on OKT8+.	Yuan et al, China, 1993
367 cases of malignant tumors	Pn chemotherapy-induced leukocytopenia.	Acupuncture and moxibustion with warming needles (A); moxibustion with ignited moxacone (B).	Effective rate was 88.4% for A and 90.9% for B. The total effective rate was 38.2% when compared with control group using batylalcohol and pentoxyl ($p<0.01$). Suggested teatment raised the effect on white cells, having no relevance to the kind of disease, or the chemotherapy.	Chen et al, Zhengzhou China, 1992
100 patients	Antiemetic in cancer chemotherapy.	Patients whose chemotherapy-induced sickness was not adequately controlled.	Best resutls were obtained from the 2 hourly self-administration of 5 min of transcutaneous electrical stimulation of P6 using a simple battery-operated TENS machine at 15 Hz, increasing the current until Qi was elicited.	Dundee et al, U of Belfast, N. Ireland, 1991

Table 3.2.4. Effect of Acupuncture on Cancer Patients (cont.)

Subject	Objective	Method	Results	Authors
26 women treated with cisplatin chemotherapy	Combined drugs with acupuncture.	A combination of metoclopramode, dexmathasone and diphenhydramine for anti-emetic treatment with acupuncture.	Acupuncture was shown to increase complete protection from nausea and to decrease the intensity and duration of nausea and vomiting.	Aglietti et al, Italy, 1990
141 patients	On edema from radiation or combination therapy of breast and uterus cancer.	Acupuncture was given to sufferers from radiation injuries to skin and soft tissues.	Radionuclide and rheographic studies as well as evaluation of hemostatic function showed acupuncture to be effective for edema and pain. It also improved lymph flow, rheovasographic indexes, and normalized hemostasis. The best results were obtained in cases of state I-II edema.	Bardychev et al, Russia, 1988

Table 3.2.5. Effect of Acupuncture on Diabetes Mellitus

Subject	Objective	Method	Results	Authors
A. Animals				
rabbits	EA at EX-B3 and ST36	Diabetes induced by Alloxan; measured blood glucose (BG), and pancreatic glucagon.	EA at EX-B3 significantly lowered the BG content and release of PG. Together with ST36, effects are even more obvious. Suggested synergetic effect from two acupoints.	Zeng et al, Chengdu, China
rats	On streptozotocin-induced diabetes.	On proliferation and expression of neuropeptide Y (NPY) in dentate gyrus (DG).	Stimulation of the ST 36 point resulted in increased cell proliferation and neuropeptide Y levels in the diabetic group.	Kim et al, South Korea, 2002
diabetic rats	Hypoglycemia induced by EA at Zhongwan.	Plasma concentrations of insulin, glucagon and beta-endorphin were determined by radioimmunoassay.	Suggested that EA at the Zhongwan acupoint induces secretion of endogenous beta-endorphin, which reduces plasma concentration in an insulin-dependent manner.	Chang et al, Taichung, Taiwan, 1999

Table 3.2.5. Effect of Acupuncture on Diabetes Mellitus (cont.)

Subject	Objective	Method	Results	Authors
diabetic rats	EA versus trans-cutaneous electric nerve stimulation (TENS).	EA or TENS at bilateral Shensu and ST36; glucose level.	EA lowered the increased level of plasma glucose ($p<0.05$); symptoms of polyphagia, polydipsia and polyuria were attenuated. Motor nerve conduction velocity slowing was prevented. Elevated the lowered pain threshold; ingeneal EA was better than TENS.	Mo et al, Beijing, 1996
B. Patients				
15 patients with type II diabetes	On gastric dysrthymia.	EA at ST36 (2Hz); cutaneous electrogastrography was performed; serum gastrin, motilin, and pancreatic poly-peptide were measured.	There was a significant increase in the percentages of normal frequencies during and after acupuncture ($p<0.01$). The percentage of tachygastric frequency was decreased significantly. Acupuncture may enhance regularity of gastric myoelectrical activity.	Chang et al, Taichung, Taiwan, 2001
diabetic patients	Laser therapy.		Laser therapy promotes compensation; has antiatherogenic, antioxidant, immunomodulating effects, improved microcirculation, myocardial contractility and performance capability.	Bodnar et al, Ukraine, Russia, 1999

Table 3.2.5. Effect of Acupuncture on Diabetes Mellitus (cont.)

Subject	Objective	Method	Results	Authors
46 diabetic patients	Chronic peripheral diabetic neuropathy.	Six courses of acupuncture analgesia for ten weeks.	77% showed significant improvement in their primary and/or secondary symptoms (p<0.01). Suggests that acupuncture is a safe and effective therapy for long-term management of painful diabetic neuropathy.	Abuaisha et al, U of Manchester, UK, 1998
patients with diabetes	On diabetic angiopathies of lower extremities.	Laseropuncture.	It had a pronounced clinical effect: removing the pain syndrome, improvement of peripheral circulation, extremity function and function of the lower extremities; improvement of thermographic values, obliterating atherosclerosis of the legs.	Peshko et al, Russia, 1992
55 patients; insulin-dependent	Diabetic angiopathy of lower extremities.	A course of 10 sessions; rheovasography, thermography and ultrasound dopplerography were performed.	A direct noticeable clinical effect was obtained in 78.2% of cases, determined by improved elastotonic properties of arteries of average caliber, enhanced blood outflow, and regulation of lower limb vascular peripheral resistance.	Solun et al, Russia, 1991

Table 3.2.6. Effect of Acupuncture on Headache

Subject	Objective	Method	Results	Authors
26 trials, 1151 patients; idiopathic headaches	Is acupuncture more effective than placebo?	Randomized and quasi-randomized clinical trials on treatment; quantitative meta-analysis not possible.	Overall the existing evidence supports the value of acupuncture for the treatment of idiopathic headaches.	Melchart et al, Munich, Germany, 2001
700 patients	EA on cranio-facial pain	20 minutes treatment with application numbers ranging between 10 and 20.	Acupuncture and infrared laser reflex therapy represent a harmless and effective treatment of such a diffuse and invalidating disease.	Costantini et al, Rome, Italy, 1997
45 patients with facial pain or headache	Acupuncture, occlusal splint and control.	Patients are randomly selected into these three groups.	Both acupuncture and occlusal splint therapy significantly reduced subjective symptoms and clinical signs from the stomatognathic system. It is concluded that acupuncture is an alternative to traditional methods.	Johansson et al, U of Gothenberg, Sweden, 1991
14 patients with tension headaches	A controlled single case design time series.	Eight weekly treatments, four true acupuncture and four shams.	True acupuncture was shown to be significantly superior to sham, demonstrating specific therapeutic action in four patients.	Vincent, London, 1990

Table 3.2.6. Effect of Acupuncture on Headache (cont.)

Subject	Objective	Method	Results	Authors
348 patients	Chronic pain syndromes.	Finnish survey: mean number of primary series: 5 sessions. 41% of patients received more than one series.	Analysis showed significant relief of pain (more than 40% reduction on the visual analog scale) in myofascial syndromes affecting the head, neck, shoulder and arm.	Junnilla, Finland, 1987
18 patients; chronic tension headaches	Controlled trial on patients with mean disease duration of 15 years.	All patients suffered from daily recurring headache, intensity recorded for 15 weeks. Crossover and randomization used.	Acupuncture was found to be significantly more pain-relieving than placebo, according to the pain registration of the patients themselves. The pain reduction was 31%.	Hansen et al, 1985

Table 3.2.7. Effect of Acupuncture on Neck Pain

Subject	Objective	Method	Results	Authors
117 patients; chronic neck pain	Comparison of acupuncture with massage.	Maximum pain related to motion (visual analog scale); 3D ultrasound analyzer, pressure algometer; quality of life.	Acupuncture showed significantly greater improvement in motion compared with massage (p<0.0052); especially better for those with myofascial pain syndrome longer than five years.	Irnich et al, Munich, Germany, 2001
70 patients; chronic neck pain	Comparison with physiotherapy.	Measured by visual analogue scale.	Both treatment groups improved in all criteria; acupuncture was more effective with patients with higher baseline pain scores.	David et al, U of Reading, UK, 1998
46 patients; chronic myofascial neck pain	Japanese acupuncture.	Randomized single-blind trial; acupuncture versus nonsteroidal anti-inflammatory medication; questionnaire, health survey, physiological measurement.	Relevant acupuncture with heat contributes to modest pain reduction in persons with myofascial neck pain; alternative therapy is recommended for the future.	Birch et al, Netherlands 1998
30 patients, cervical spine pain	Randomized controlled study on patients suffering with a mean duration of 8 years.	12 weeks of treatment.	80% of treated group felt improved, with a mean of 40% reduction of pain score, 54% reduction of pain pills, 68% reduction of pain hours per day, and 32% less limitation of activity. Only 13% of controlled felt better.	Coan et al, 1981

Table 3.2.8. Effect of Acupuncture on Shoulder Pain

Subject	Objective	Method	Results	Authors
44 patients, shoulder's myofascial pain	Comparison between superficial and deep acupuncture.	Treatment with 13 needles; 4 trigger points in the shoulder area; 8 sessions.	Both techniques had efficacy in controlling pain. Deep acupuncture shows to be better at all times with significantly better results after treatment and follow-up after one and three months.	Ceccherelli, et al, Italy
35 patients; frozen shoulder	On treatment effectiveness.	Functional mobility, power and pain at 6 and 20 weeks. Assessed by constant shoulder assessment.	Improvements in scores by 39.6% and 76.4% were seen, from exercise and exercise plus acupuncture, respectively; scores became 40.3% and 77.2% at 20 weeks (p<0.025).	Sun et al, Hong Kong, 2001
52 sportsmen	On rotator cuff tendonitis.	Modified Constant-Murley score; randomized single blind; new placebo-needle.	The acupuncture group improved 19.2 points, while the control group improved 8.37 points (p=0.014).	Kleinhenz et al, U of Germany, 1999

Table 3.2.8. Effect of Acupuncture on Shoulder Pain (cont.)

Subject	Objective	Method	Results	Authors
83 cases; postthermiplegic omalgia	Pathogenesis of shoulder pain.	Large range of passive movement leasing to omalgia.	Suggests that painless movement of the shoulder joint should be limited; massage should be given immeditely after acupuncture.	Chen et al, Fuzhou, China, 1998
40 cases, shoulder-hand syndrome	Treatment with EA.	Comparison with filiform needle acupuncture (FNA).	EA had better results in treating hand back swelling, hand skin temperature elevating, and the finger pain caused by bending than that with FNA (p<0.05). Also better in shoulder joint improvement. Total effect rate of EA is 75% versus 50% in FNA. EA produced a rhythmic muscle contraction.	Guo et al, Hefei, China, 1995
150 patients; frozen shoulder	EA plus nerve block.	Pain relief from EA and regional nerve block by 1%; xylocaine; six vector movements were checked.	Results showed that the combined EA and nerve block had significantly high pain control quality, longer duration, and better range of movement of shoulder joint than that of EA or nerve block alone.	Lin et al, Taipei, China

Table 3.2.8. Effect of Acupuncture on Shoulder Pain (cont.)

Subject	Objective	Method	Results	Authors
37 patients	On chronic neck and shoulder pain.	All patients had been unresponsive to conventional or placebo treatments; double blind evaluation was used.	64.9% patients obtained significant long term improvement. An increase in microcirculation might be responsible for the tissue healing and subsequent relief.	Fang et al, Hangzhou, China, 1987
971 cases	Acupuncture for different diseases	It was regarded as successful if patients felt no pain and had significant improvement.	Positive results were obtained for cephalagias, sinusitis, cervical spine syndrome, shoulder-arm syndrome, back pain, constipation, ischialgias, herpes zoster, allertic rhinitis and disturbances of peripheral blood flow.	Fisher et al, 1984

Table 3.2.9. Effect of Acupuncture on Low Back Pain

Subject	Objective	Method	Results	Authors
131 patients	Chronic low back pain (LBP); suratio of pain: 9.6 years	Randomized, blinded placebo-control trial; 12 weeks of physiotherapy and then control trial; 12 weeks of physiotherapy and then 20 sessions of acupuncture.	Acupuncture was superior to control regarding pain intensity (p=0.00), pain disability (p=0.00), intensity (p=0.00). The trial also demonstrated that there was a placebo effect.	Leibing et al, Goettingen, Germany, 2002
60 patients	Low back pain and radicular symptoms.	Lumbar disc herniation from magnetic resonance imaging; pain intensity assessed by visual analog scale.	Intensity of low back pain dropped from 59 to 19 mm, and intensity of radicular pain from 64 to 12 mm after treatment. 3 to 12 months later 88% were satisfied with the treatment.	Schmitt et al, Heidelberg, Germany, 2001
75 elderly patients	Low back pain and knee pain.	A cross-sectional study was carried out in a geriatric hospital.	60 patients answered that their pain diminished following acupuncture (p=0.05); however, 46% of them were also teated with other types of physical therapy.	Washio et al, Japan, 2001
60 patients; acute low back pain	Acupuncture versus antiplogistica.	Randomized into two groups: one received acupuncture, the other received naproxen.	Patients receiving acupuncture used significantly less analgesic drugs; reported fewer new episodes of low back pain (p<0.01).	Kittang et al, Norway, 2001

Table 3.2.9. Effect of Acupuncture on Low Back Pain (cont.)

Subject	Objective	Method	Results	Authors
109 patients	Acupuncture massage versus Swedish massage.	Randomized controlled 2 x 2 factorial design; measured functional disability, pain intensity, pre/post changes.	Acupuncture massage showed beneficial effects for both disability and pain, compared with Swedish massage ($p=0.003$ and 0.024 respectively).	Franke et al, 2000
62 cases; 3rd lumbar tansverse	EA at Huatuohianji versus conventional acupuncture.	32 with EA at Huatuohianji; and 30 cases by conventional acupuncture.	EA at Huatuohianji points was superior in analgesic effect and clinical effective rate; possibly related to the trunk of posterior ramus of the spinal nerve where the points are located.	Wang et al, Guangzhou, China, 1999
24 patients	Low back pain after removal of nucleus pulposus.	Treated with silver needle at tender points in low back and buttock.	Pain scores for each tender point after treatment were significantly lower than those taken before treatment ($p<0.001$).	Yi-Kai et al, Guangzhou, China, 2000
60 pregnant women	Acupuncture versus physiotherapy for low back and pelvic pain.	Measured pain with visual analog scale, and disability by disability rating index (DRI).	Pain scores reduced from 3.7 to 0.9 in the morning and from 7.4 to 1.7 ($p<0.01$ for both values) in the evening for the acupuncture group, and DRI also decreased significantly. Both scores showed improvements which were much better than the physiotherapy group.	Wedeberg et al, Sweden, 2000

Table 3.2.9. Effect of Acupuncture on Low Back Pain (cont.)

Subject	Objective	Method	Results	Author
20 patients with low back pain	Randomized controlled trial.	Measured by disability questionnaire, pain scale, daily living scale and SF-36 general health questionnaire.	Post treatment and 6 months after the end of treatment. There were statistically significant improvements in disability, pain intensity, daily living, physical functioning, social functioning, bodily pain, vitality and mental health.	MacPherson et al, York, UK, 1999
a 29 year old man; several years of back pain	Bone metabolism.	High resolution bone scan of the skeleton; patient received acupuncture treatment.	Posterior and lateral images showed focal increased intake in several regions of the skull; hence acupuncture can cause enhanced bone metabolism.	Kuno et al, Seattle, 1995
56 cases	Disturbance in small articulation of the lumbar vertebrae.	Puncturing the effective points and reduction by manual traction.	The cure rate was 39.3%, the markedly effective rate 28%, with a total effect rate of 98.3%.	Zhang et al, Beijing, China, 1994
40 patients; low back pain	Modes of acupuncture.	Three modes of acupuncture: manual, EA at 2 Hz and at 80 Hz; evaluated at 6 weeks and 6 months after treatment.	After 6 months patients receiving 2 Hz treatment showed significant improvement, whereas patients receiving manual, or 80 Hz did not. Suggested 2 Hz is the mode of choice.	Thomas et al, Sweden, 1994

Table 3.3.1. Symmetry Effect of Acupuncture

Subject	Objective	Method	Results	Authors
54 patients	Effect of acupuncture in accident patients.	Using skin temperature.	There was a very significant rise in skin tempertaure in the area of lesion, which persisted beyond the teatment period. The effect was also observed on the contralateral untreated extremity.	Rabi et al, Germany, 1983
65 cases, rats	Analgesic effect of ipsilateral (IS) and controlateral (CS) stimulation.	Electro-acupuncture (EA); testing rats' pain threshold and recording the neuronal activity in the D-PAG.	It was verified that IS and CS had similar effects on pain relieving. Neither IS nor CSs could increase pain threshold in the unilateral D-PAG lesioned rats. The excited neuronal discharge was recorded in the unilateral D-PAG by stimulating rats' Zusanli ST36 at each side.	Fang et al, Hangzhou, 1994

Table 3.5.1. Exponential Decay

Patients	Terminal temperature a	Temperature decrease b	Time Decay constant (min)	Chi Squared per degree of freedom	Acupoint	Measured Pain Area
1	35.06	1.16	12.6	0.00639	GB34	right buttock
2	35.29	0.967	7.87	0.0167	GB34	right buttock
3	34.3	0.769	4.5	0.016	DU14	left shoulder
4	33.05	2.95	22.59	0.0242	BL60	left popliteal area (behind the knee)
5	31.36	3.75	29.52	0.0189	BL60	right popliteal area
6	34.02	0.712	6.348	0.0432	SI3	shoulder
7	35.99	1.501	6.28	0.029	LU5	fist area
8	35.9	1.553	7.47	0.00651	LU5	fist area
9	36.12	0.59	4.3	0.0186	LU5	wrist
10	35.54	3.89	25.26	0.00924	BL40	right shoulder
11	31.37	8.29	63.58	0.0475	BL40	left arm
12	37.81	2.929	20.14	0.0166	BL40	right neck
13	35.27	5.57	32.04	0.0308	BL40	neck, left side
14	36.64	0.396	10.58	0.0193	DU14	neck, right side
15	34.39	1.254	9.64	0.0228	SI3	back
total	522.11	36.281	262.718			
averages	34.8073333333	2.4187333333	17.5145333333			

Table 4.1.1. Effect of acupuncture on obesity

Subject	Objective	Method	Results	Authors
obese rats	Monoamines and adrenosine triphosphatase in hypothalamic area (LAH).	Central nerve high-pull perfusion and biochemical techniques.	In obese rats, the levels of noradrenaline (NA) were higher, serotonin (5-HT) lower, and activity of ATPase in LHA lower. After acupuncture, levels of NA reduced and 5-HT and activity of ATPase increased, weight of obese rats reduced. Anti-obesity effect of acupuncture might come from effective regulation of LHA.	Liu et al, Nanjing, 2000
202 children with simple obesity	Effects of self-made photo-acupuncture.	Obesity indices.	Obesity indices were lowered significantly, and levels of blood lipids, glucose, cortisol and triiodothyronine were all improved markedly. Photo-acupuncture is a safe, painless, nontraumatic and effective method for treatment of simple obesity, and easily acceptable.	Yu et al, Changchun, 1998

Table 4.1.1. Effect of acupuncture on obesity (cont.)

Subject	Objective	Method	Results	Authors
rats	On satiety center of ventromedial nucleus of hypothalamus.	Microelectrode recording method of nerve cells, and steriotaxic techniques were adopted; discharge frequency of nerve impulse in VMH as the index.	Electric activity in acupuncture group was higher than in the obesity group ($p<.001$) and the normal group ($p<0.01$). Acupuncture can increase excitability of the satiety center with a better long term effect.	Zhao et al, Nanjing, 2000
359 adult females	Keep shape for non-obese females.	Combined application of body acupuncture, moxibustion and auricular acupuncture.	Very effectively regulates the somatotypic indices of body weight, circumference of the chest, loin, hip and thigh, the ratio of the loin, sebum thickness, obesity degree, body mass index and body fat percentage. Concluded this is very good therapy.	Liu et al, Nanjing, 1998
60 overweight subjects	Weight loss by auricular acupuncture.	Attached AcuSlim to the ear points, shenmen and stomach, for four weeks.	95% noticed suppression of appetite. Both the number of subjects who lost weight by 2 kg and the mean weight loss were significantly higher ($p<.05$).	Marley Adelaide, Australia, 1998

Table 4.1.1. Effect of acupuncture on obesity (cont.)

Subject	Object	Method	Results	Authors
34 obese patients	Effect on obese patients with hyperlipidemia.	Obesity indices, lipid index (TC, TG, LDL-C, HDL-C), atherosclerosis index (AI), ratio of waist to hip (W/C), TXB2 and 6-keto-PGF1 alpha.	Showed marked weight loss effect, while the levels of TC, TG, LDL-C, HDL-C, AI, W/H, TXB2, 6-keto-PGF-1 alpha were finely regulated. Suggested acupuncture treated obesity and hyperlipidemia and resisted the pathological factors leading up to circular diseases.	Liu et al, Nanjing, 1996
75 patients	Anti-obesity effect and influence on water and salt metabolism.	Observed changes in obesity indices, blood sodium, blood potassium, mOsm of plasma and urinary aldosterone before and after acupuncture.	Acupuncture changed the blood sodium, blood potassium aldosterone, and mOsm plasma to normal levels. Indicated acupuncture has anti-obesity effect and improved the water and salt metabolism of patients by regulation of nervous system and body fluid.	Sun, Nanjing, 1996
45 cases: 8 males and 37 females	Triple therapy for obesity.	Arucular acupuncture, diet control and aerobic exercise for eight weeks.	The rate of effectiveness was 86.7%. The rate of body weight rebound (regained more than 1.5kg) was 6.7% and 18.9% one month and one year later.	Huang et al, Kaoshiung, Taiwan, 1996

Table 4.1.1. Effect of acupuncture on obesity (cont.)

Subject	Objective	Method	Results	Authors
35 healthy non-obese subjects	Effects of auricular acupuncture.	Small needles were applied intracutaneously into the bilateral cavum conchae, that has resistance of 100k omega.	57.1% reduced their weight by charting their body weight four times a day alone. With aricular acupuncture 70.4% reported reduction of weight.	Shiraishi et al, Tokai U, Japan, 1995
718 patients with simple obesity	On patients with stomach-intestine excessive heat type.	Obesity index and biochemical indices before and after acupuncture were observed.	Marked weight loss effects were achieved, while biochemical indices improved. Suggests good regulatory effect on the function of nerve, endocrine, digestion and energy metabolism.	Liu et al, Nanjing, 1995
obese rats	On feeding-related hypothalamic neuronal activity (LHA)	While stimulating low resistance ipsilateral vagal innervated region, recording was via a glass microelectrode.	The latency of potential evoked in LHA was 28.1+/-3.3 ms; LHA neuronal activity was depressed 45.6%; ventromedial hypothalamic neuronal activity was excited. Suggested satiation formation and preservation.	Shiraishi et al, Tokai U, Japan, 1995

Table 4.1.1. Effect of acupuncture on obesity (cont.)

Subject	Objective	Method	Results	Authors
19 obese patients	To reduce weight	Electro-acupuncture	Patients with knee osteoarthritis participated in EA, diet, and exercise program. 89% of them lost 5-10 kg.	Shafshak, Egypt, 1995
rats	Activation of the satiety center.	Stimulation of inner auricular regions that correspond to human pylorus, lung, trachea, stomach, esophagus and heart.	After rat gained 410 g in 20 days, acupuncture reduced it to its original 290 g. Acupuncture evoked potential in hypothalamic ventromedial nucleus (HVM).	Asamoto et al, Showa U, Tokyo, 1992
102 cases	On simple obesity complicated by cardiovascular diseases.	Changes of pathogenetic and hazardous factors and regulating function of vegetative nerves on cardiovascular activities.	Total effective rate of 88.24%; benign conversion effect on loin/hip ratio, the arteriosclerotic index, and the vegetative nervous system. This is an important method for preventing and treating complicating cardiovascular diseases.	Liu et al, Nanjing, 1992
46 cases of simple obesity	Anti-obesity plasma cyclic nucleotide and nervous system.	Changes of obesity index, lipid index, equilibrium indices of the vegetative nervous system (Y), the plasma cAMP.	Effective rate was 84.8%. It also brought about biphase change on the blood pressure and Y values. Good regulative effect on lipid metabolism and plasma cAMP of patients.	Liu et al, Nanjing, 1991

Table 4.1.1. Effect of acupuncture on obesity (cont.)

Subject	Objective	Method	Results	Authors
39 cases of simple obesity	Hypothalamus-pituitary-adrenal axis.	Observed obesity indices, lipid level, content of ACTH in plasma and salivary cortisol before and after acupuncture.	Markedly effective rate was 38.5%; the effective rate was 51.3%; total effective rate was 89.7%. Regulated the lipid level and enhanced the function of hypothalamus-pituitary-adrenal system.	Liu, ZC, Nanjing, 1990
41 obese patients with hypertension	Effect of acupuncture and moxibustion.	Observed the obesity indices, lipid indices (TC, TG, VLDL-C, TC/HDL-C, HDL-C, LDL-C, LDL-C/HDL-C and AI) and physiological indices (saliva secretion, heart rate, respiratory rate, blood pressure, and temperature), energy metabolism indices (BMR).	Total effective rate was 87.8% (36 cases); showed good therapeutic effect; benign regulatory effect in overeating, blood pressure, the vegetative nervous indices, the lipid level and the energy metabolism.	Liu, ZC, Nanjing, 1990
196 cases of simple obesity	On high-density lipoprotein cholesterol.	Changes in obesity indices and lipid indices.	Good therapeutic effect on obesity was obtained. Also benign regulatory effect on lipid metabolism and high density lipoprotein cholesterol.	Liu, ZC, Nanjing, 1990

Table 4.1.2. Effect of acupuncture on aging

Subject	Objective	Method	Results	Authors
rats	On aging process of genital system.	Catgut embedding at BL23; EA at BL23.	Catgut embedding shortens sexual cycles, incrases the frequency of sexual cycle, and slows down the aging process of the genital system in both the aged rats and rats with injured noradrenergic endings. After EA, frequency of neuronal discharges in locus coeruleus (LC) was elevated, and the activating rate of LC to neurons in the medial preoptic area of hypothalamus was increased. Suggested it raised the catecholamine (CA)/5-hydroxytryptamine (5-HT) ratio in the hypothalamus so as to delay the aging process	Zhu et al, Shaanxi, 2000

Table 4.1.2. Effect of acupuncture on aging (cont.)

Subject	Objective	Method	Results	Authors
ovariect-omized mice	On catecholamine content in brain; memory loss.	Specific brain regions were assayed for catecholamine contents by liquid chromatog-raphy with electro-chemical detector. The mitogenic activities of splenic lymphocytes were measured. Learning and memory ability were studied by step-through type passive avoidance test.	EA increased norepinephrine and dopamine contents in brain region and enhanced mitogenic activities of splenic lymphocytes. It also improved memory related behavior. Overall changes were observed in central nervous system (including retention of memory) and immune functions. Reduced memory loss and improved decrease of immune responses accompanying aging and/or menopause.	Torizuka et al, Tokyo, 1999
normal subjects	On the amount of telomere molecules.	Use synthesized basic units of human telomere molecules as reference control.	Found that acupuncture on ST36 on normal subjects increased the telomere levels up to a maximum of 2 times their telomere levels before treatment. Frequently increases were between 60% to 100%. Strong shiatsu performed on ST36 produced a somewhat lesser effect than acupuncture.	Omura et al, New York, 1998

Table 4.2.1. Effect of qigong on healthy persons

Subjects	Method	Results	Authors
eight fighter pilots	Mechanism of raising the blood pressure by Qigong (QG); manuever at +1 G.	During Q-G maneuvering, even with a high-G load, the thoracic pressure remained negative or at low pressures, while gastric pressures were remarkably raised. A relatively large and constant pressure gradient between abdominal and thoracic pressures were maintained.	Zhang et al, Beijing, 1992
18 fighter pilots	Qigong maneuver; tests at +1Gz, and centrifuge tests; blood pressure.	Tolerance to Rapid Onset Rate G-load is 3.82 G in normal conditions. After QG, it rose to 6.64 G, a gain of 2.82 G. QG does not lead to hypoxia or hyperventilation. It is an innovative G-protective maneuver.	Zhang et al, Beijing, 1991
11 healthy adults	He Xiang Zhuang Gongfu; six months' practice.	CO_2 decreased after 2 and 6 months' practice, indicating reduction of metabolic rate and decreased consumption of oxygen. MVV has an up tendency, indicating strengthening respiratory muscle. Exercise can induce a wakeful hypometabolic physiologic state.	Yan et al, Sichuan, 1993
24 subjects; zen meditation	Frontal midline theta rhythm (Fm theta); simultaneous measurement of EEG, ECG.	Close relationship between cardiac autonomic function and activity of medial frontal neural circuitry.	Kubota et al, Kyoto U, 2001
19 volunteers	Changes of cytokine production; immunological function.	IL4 and IL2 remained stable; IL6 increased; IFN gamma increased and TNF alpha increased; IL10 decreased. Cortisol, known inhibitor of type 1 cytokine production, was reduced. Blood levels of stress-related hormone cortisol were lowered. Concommitant changes in numbers of cytokine-secreting cells.	Jones et al, Hong Kong, 2001

Table 4.2.1. Effect of qigong on healthy persons

Subjects	Method	Results	Authors
7 healthy subjects	Electroacupuncture Voll; before and after qigong exercise.	QG exercise changed the average EAV measured values at the rate of -17% to -35% for four subjects, and 4 to 15% for three subjects.	Sancier, Menlo Park, 1994
Qigong trainee	Skin tests for delayed cutaneous hypersensitivity, with seven antigens.	The maximal antigen response time (24 hr) was faster and the response antigen number (6) was higher than control group (48 hr and 4 antigens). This represents the difference in cell-mediated immunity between Qigong tainees and normal subjects.	Ryu et al, Korea, 1995
26 healthy volunteers	R-R interval spectral analysis variability of heart rate.	Breathing pattern A, B, C: amplitude of peak in high frequency increased with a reduction of LF2/HF ratio. Breathing pattern D: amplitude of peak decreased with an increase of LF2/HF ratio. Breathing pattern D: amplitude of peak decreased with an increase of LF2/HF ratio. QG could indirectly regulate the function of viscera after controlling the direct breathing pattern.	Sun et al, Beijing, 1992
37 subjects exercised in Neiyang Gong	Flash visual evoked potentials recorded from the occipital scalp.	QG mediation may have either facilitative or inhibitory effects on the visual cortex depending on QG methods.	Zhang et al, Shanghai, 1993

Table 4.2.1. Effect of qigong on healthy persons

Subject	Method	Results	Authors
12 healthy trainees	Chun DoSunBup (CDSB) Qi training.	Heart rate, respiratory rate, systolic blood pressure and rate-pressure product were significantly decreased during Qi-training. Indicates stabilization of cardiovascular system.	Lee et al, Korea, 2000
Elderly group and youth group of CDSB (Chun-DoSunBup) trainees	Response of plasma growth hormone GH, insulin-like growth factor-I (IGF-I); testosterone T.	Plasma GH levels increased 7.26-fold in the elderly and 1.66-fold in the young. IGF-I levels increased significantly in the young but not in the elderly. The T level increased significantly in elderly subjects but not in the young. Hence, CDSB may be the therapy applicable to growth factor-related disorders such as GH deficiency in children and osteoporosis in the elderly.	Lee et al, Wonk-wang U, Iksan, Korea, 1999

Table 4.2.2. Effect of Qigong Therapy

Subjects	Method	Results	Authors
Patients with heart qi deficiency and blood stasis type, treatd with qigong	Qigong practice to treat hypertension. Duration: one year.	Cardiac morphology and function, hemorheology and erythrocyte deformity were improved. Plasma histofibrinogen activaton inhibitor (PAI), VIII factor-related antigen levels decreased. Plasma tissue fibrinolytic activator and anti-thrombogen III levels increased. Capillary blood velocity of nailfold microcirculation raised from 0.2940+/-0.0206 mm/s to 0.3045+/-0.0236 mm/s. Diameter and length of efferent limb tended to increase.	Wang et al, Shanghai, 1995
22 treatment-resistant patients with late-stage complex regional pain syndrome type I.	Qigong and qi emission by qigong master; thermograph, swelling, discoloration, muscle wasting, frequency of pain awakening; mood, anxiety assessment.	Qigong training results in transient pain reduction and long-term anxiety reduction. After first session 82% reported less pain, as against 45% in control group. After 3 weeks, 91% of patients reported analgesia, as against 36% in control group.	Wu et al, Newark, 1999
Ten in-patients with diabetes mellitus.	Qigong walking (30-40 min).	Plasma glucose levels decreased from 223 to 216 mg/dL, and the heartbeat from 70 to 79 beats per min. QG reduced plasma glucose without inducing a large increase in pulse rate.	Iwao et al, Kyoto, 1999

Table 4.2.2. Effect of Qigong Therapy (cont.)

Subjects	Method	Results	Authors
10 cases of patients with aplastic anemia.	Qigong therapy changes in T-cell: helper T cell (Th), suppressor T cell (Ts).	Ratio of Th/Ts was greatly elevated. Ts went down, but not significantly.	Yao, 1989
Qi Gong master, patient.	Bi-digital O-Ring.	Acupoints RN5 and RN6 show changes from +4 in the pre-qigong state. Similar changes are abserved in RN17, RN22, DU20, entire pericardium and triple beat meridians, etc.	Omura et al, New York, 1989
68 subjects with cardiovascular disease, joint diseases, respiratory and other diseases.	Fluorescence spectrophotometry to observe the variation of blood contents of monoamine neurotransmitter 5-hydroxytamine (5-HT), norepinephrine (NE) and dipamine (DA).	Comparison pre- and post- exercise showed a reduction in 5-HT from 0.43+/-0.21 to 0.21+/0.13 microgram/ml; NE from 0.27+/-.13 to -.35+/-0.27; DA from 0.86+/-0.69 to 1.19+/-0.81 microgram/ml. The post-exercise blood content of DA in various groups rose remarkably.	Liu et al, Hefei, 1990
45 patients of essential hypertension.	Qi-gong therapy.	Plasma 6-K-PGF1 alpha was increased, and TXB2 was decreased after therapy. Suggests qigong is regulatory on TXB2 and 6-K-PGF1 alpha in patients with essential hypertension.	Li et al, Shangsha, 1997

Table 4.2.2. Effect of Qigong Therapy (cont.)

Subjects	Method	Results	Authors
Patients	Qigong therapy.	Physiological effects: changes in EEG, EMG, respiratory movement, heart rate, skin potential, skin temperature and finger tip volume; sympathetic nerve function, function in stomach and intestine; metabolism; endocrine and immune system. Psychological effects: motor phenomena and perceptual changes; experienced warmth, chilliness, itching sensation, numbness, soreness, bloatedness, relaxation, tenseness, floating, dropping, enlargement or constriction of the body image, etc.	Xu et al, Shanghai et al, 1994
60 cases of pregnancy-induced hypertension.	Qi-gong relaxation exercise.	(1) Evaluation according to PIH combined score, effective in qigong group, 54 out of 60; control, 33 out of 60. (2) Meconium stain in amniotic fluid present in 20, 0% in qigong gorup, and 48.3% in control group. (3) Incidence of abnormal hematocrit decreased from 52.4% to 23.8% after treatment. (4) The mean value of blood E2 by RIA showed increase from 22.97+/-13.16 microgm/ml to 33.75+/-34.01 after treatment.	Zhou et al, 1989

Table 4.3.1 Effect of auricular acupuncture

Subjects	Objective	Method	Results	Authors
A. Human Subjects				
volunteers: 10 men and 10 women	Auricular electrically stimulated analgesia.	Randomized, double blind, corssover trial, twice, with desflurane, 10mA at 299 Hz and 149 Hz.	Electrical stimulation of the lateralization control point reduced anesthetic requirement by 11+/-7% (p<0.001), with the reduction being similar in women and men.	Greif et al, UC San Francisco, 2002
55 volunteers	Treatment for anxiety.	Auricular acupuncture at a relaxation point.	Patients in the relaxation group were significantly less anxious at 30 min (p=0.007) and 24 h (p=0.035), as compared to control groups. Auricular acupuncture at the relaxation point can decrease the anxiety level in a population of healthy volunteers.	Wang et al, Yale U, USA, 2001
10 volunteers	Reduction of anesthesia requirement by auricular acupuncture.	On anesthesia with desflurane needles placed at Shen Men, thalamus; tranquilizer and cerebral points on right ear.	Auricular acupuncture reduced anesthetic requirement by 8.5%.	Tahuchi et al, 20002

Table 4.3.1. Effect of auricular acupuncture (cont.)

Subjects	Objective	Method	Results	Authors
23 healthy volunteers	On olfactory acuity.	Controlled single blind, randomized study. The lung point was employed; two odorous beta-phenyl ethyl alcohol and methyl cyclopentenolone were used.	A significant decrease in the olfactory recognition threshold by acupuncture in comparison with controls ($p<0.05$) for two standard odors used.	Tanaka et al, Fukuoka U, Japan, 1999
60 overweight subjects	On weight loss.	Used the AcuSlim device twice daily for four weeks on Shenmen and stomach points.	95% of the treated group noticed suppression of appetite, whereas none of the control group noticed such change. Both the number who lost weight, and the mean weight loss were significantly higher in the treated group ($p<0.05$).	Richards et al, U of Adelaide.
45 infertile women	Treatment of infertility.	Auricular acupuncture versus hormone teatment.	Women treated with acupuncture had 22 pregnancies, 11 after acupuncture, whereas women treated with hormones had 20 pregnancies. Suggests auricular acupuncture as a valuable alternative therapy for female infertility due to hormone disorders.	Gerhard et al, U of Heidelberg, Germany, 1992

Table 4.3.1. Effect of auricular acupuncture (cont.)

Subjects	Objective	Method	Results	Authors
30 patients	On blood pressure and cardiac function.	Comparison of heart point versus stomach point in the ear.	Blood pressure was lowered; there was a marked improved effect of left cardiac function, with II, III stage of hypertension from treatment at the heart point, much less so at the stomach point.	Huang et al, Wuhan, China, 1991

B. Animal Studies

Subjects	Objective	Method	Results	Authors
food-deprived rats	Decrease of neuropeptide Y expression.	Needling the unfed rats on the auricular points; immunochisto-chemistry was used to detect enhanced NPY.	Present findings indicate that auricular acupuncture may affect NPY expression in the arcuate nucleus (ARNO) and the paraventricular nucleus (PVN) of the hypothalamus.	Kim et al, Semyung U, South Korea, 2001
rats	Cell proliferation in the dentate gyrus.	Use 5-Bromo-2'-deoxyuridine-5'-monophosphate (BrdU) immuno-histochemistry.	It revealed a significant increase in cell birth in the dentate gyrus of both appropriately fed and food-deprived adult rats.	Kim et al, Semyung U, South Korea

Table 4.3.1. Effect of auricular acupuncture (cont.)

Subjects	Objective	Method	Results	Authors
rats	On feeding-related neuronal activity.	Stimulate the ipsilateral vagal innervated region of the auricle and record through a LHA and VMH.	Lateral hypothalamic (LHA) neuronal activity was depressed 45.6% (n=12, p<.01). Auricular stimulation modulates feeding-related hypothalamic neuronal activity of obese rats. It may not reduce appetite, but it is more likely concerned with satiation formation and preservation.	Shiraishi et al, Tokai U, Japan, 1995
rats	Activation of satiety center.	Stimulate rat's inner auricular points corresponding to lung, trachea, stomach, endocrine and heart.	Needle implantation into any of these points reduced the body weight to its initial 290 g after the rat had gained about 410 g in 20 days. Stimulation of other acupuncture points did not evoke hypothalamic ventromedial nucleus HVM potentials and did not reduce body weight.	Asamoto et al, Showa U, Japan, 1992.

Table 1.2.0. Evidence against nerves and for meridians as responsible for effects of acupuncture

Nonexistence of meridians; effects from nerves only

Method	Evidence	Data Reference	Evidence for water clusters with electric dipoles	
			Water clusters	Water
Auricular acupuncture	Lack of major nerves in ear	Table 4.3.1		
Low conductivity at acupoints	No explanation	Ch 1		
Low conductivity lines in plants	Nonexistence of nerves in plants	Ch 1		
Blocking of transmission of qi by mechanical means	No	Ch 1		
Pain relief from change of temperature at sources	No, pain relief from blocking of transmission through nerves	Ch 3		
Transmission of infrared along meridians	No	Ch 1 & Ch 3		
Bi-directional function of acupuncture	Not possible	Ch 3		
Balancing left and right as viewed by infrared imaging	No	Ch 3		

Table 4.3.2. Evidence against nerves and for meridians as responsible for effects of acupuncture (cont.)

Evidence of water clusters with electric dipoles

Method (see text)	Evidence	Data reference	Water clusters	Water
	Low impedance of meridian	Ch 1	possible	yes
	High electric potential at acupoint	Ch 4	yes	no
	Different effect from different frequency I electro-acupuncture	Ch 1	yes	no
	Excess and deficiency from same needle	Ch 3	yes	no
	Stop of propagation of sensation of acupuncture by pressing	Ch 1	yes	no
	Moxibustion as transmission of infrared radiation	Ch 1	possible	no
	Non-standard black body radiation from acupoint	Ch 4	yes	no

Table 4.4.1. Effect of External Qi on Animals

Subjects	Method	Results	Authors
rabbit and rats	External QG emitted by a quartz crystal upon application of electric current; human QG; EEG in rabbit and electrical activity of rat pineal gland.	Three type of changes are recorded in EEG similar to changes from intravenous administration of 5-hydroxytryptophan. These changes disappear after application of methysergide (10mg/kg), a serontin antagonist, or after pinealectomy. External QG might inhibit N-acetyl-transferase and increase serotonin levels.	Takeshige et al, Showa U, 1994
gastro-nemius muscle of guinea pigs	Single isometric twitch height in situ to study the pain relief mechanism.	Muscle pain relief by external qigong, acupuncture, or static magnetic field might be induced by recovery of circulation due to enhanced release of acetylcholine as a result of activation of cholinergic vasodilator nerve endings innervated to the muscle artery. Activation of cholinergic nerves might be induced by inhibition of cholinesterase, or by a somato-autonomic reflex, which might be in the anterior hypothalamus.	Takeshige et al, Showa U, 1996
tumor-bearing mice	Qigong-emitted external Qi (QEQ); murine tumor models bearing Ehrlich ascites carcinoma and ascitic sarcoma.	QEQ has inhibitory effects on tumor growth and enhancing effects on anti-tumor immunologic functions of tumor host. In combination with chemotherapeutic agent cyclophosphamide, it increases the anti-tumor efficacy, and markedly improves the compromised anti-tumor immunologic functions of the tumor.	Lei et al, Wuhan, 1991

Table 4.4.2. Physical properties of external qi by qigong masters

External qi	Measured quantities.	Frequency and strength.	Ref*
A. External Qi which are Photons			
1. Infrared	Low frequency modulation.	(1) Low frequency modulation: the measured frequencies are 0.05, 0.1, and 0.3 Hz by Master Lin; approximately 0.15 Hz by Master Tam. (2) Resonance effect reported by patient with contraction and epansion of the bladder.	p. 36
	Deviation from random black body radiation.	5%-7% in the range of 3 to 5 microns.	p. 351
2. Magnetic Field	Magnetic field in the range of 0.25 gauss; 1.67 gauss detected.	Magnetic detector placed 2 cm parallel to the skin above acupoint PC8; observed 18 peaks in 100 seconds, or 0.18 Hz.	p. 47
3. Electricity	Static electricity up to 3.4x10(-14) above acupoint EX-HN3.	Period is from 15 sec-25 sec, or average frequency of 0.05 Hz.	p. 59
B. External Qi which are Phonons			
1. Subsonic waves	Frequency and amplitude of qigong masters.	1. Frequencies are in two ranges: 1-2 Hz and 9-10 Hz. 2. Effect on the vibration of arteries of patients who received treatment from qigong master: vibration increased 530% to 690%.	p. 51
2. Subsonic waves	Sonic pressure by 10 qigong practitioners and 10 ordinary persons as control.	Sonic pressure at 1 cm above PC8: for control group. The pressure is 42.4+/-1.9dB, and that of qigong practitioner is 49.6+/-2.5 dB; p<0.001 frequency for two groups is the same, in the range of 3-5 Hz.	p. 54

*Reference: Xie W Z, Scientific Basis of Qigong, Beijing Univeristy of Science and Technology Press, 1988.

Table 5.5.1. Conservation of Number Laws

Objects	Conservation Laws	Breaking of Conservation Laws	Gauge Particle
human	Conservation of number of human beings	Death of a person	qions
organs	Conservation of number of organs	Accidents or failure of an organ	qions
cells	Conservation of number of cells	Death of a cell	qions
molecules	Conservation of number of molecules	Biochemical reactions where molecules are changed.	photons
electrons	Conservation of number of electrons	There are no experiments so far that do not conserve the electron number.	photons

Made in the USA